ANTHROPOLOGICAL PAPERS OF
THE UNIVERSITY OF ARIZONA
NUMBER 37

CERAMIC SEQUENCE IN COLIMA: CAPACHA, AN EARLY PHASE

ISABEL KELLY

THE UNIVERSITY OF ARIZONA PRESS
TUCSON, ARIZONA
1980

About the Author . . .

ISABEL KELLY received a doctorate from the University of California at Berkeley in 1932. She has done archaeological investigation in the southwestern United States and especially in West Mexico, and her ethnographic work includes California and the Great Basin (Coast Miwok, Northern and Southern Paiute, and Chemehuevi) and Mexico (Lowland and Highland Totonac, as well as several mestizo areas). In addition, she has worked in applied anthropology projects in Mexico, Bolivia, and Pakistan. A resident of Mexico for approximately four decades, since 1960 her time has been devoted to private research.

THE UNIVERSITY OF ARIZONA PRESS

Copyright 1980
The Arizona Board of Regents
All Rights Reserved
Manufactured in the U.S.A.

Library of Congress Cataloging in Publication Data

Kelly, Isabel Truesdell, 1906–
    Ceramic sequence in Colima.

    (Anthropological papers of the University of
Arizona; no. 37)
        Bibliography: p.
        1. Indians of Mexico—Colima (State)—Pottery.
2. Indians of Mexico—Colima (State)—Antiquities.
3. Colima, Mexico (State)—Antiquities.   4. Mexico
—Antiquities.   I. Title.   II. Title: Capacha,
an early phase.   III. Series: Arizona. University
Anthropological papers; no. 37.
F1219.1.C75K44   738'.0972'36   80-21301

ISBN 0-8165-0565-9

To the memory of
Carl Ortwin Sauer
and
Lorena Schowengerdt Sauer

CONTENTS

FOREWORD

The thrill of discovery comes to an investigator in archaeology when a class of objects apparently foreign to a familiar complex of cultural materials can be assigned to a place in a chronological scheme or determined to be indicators of connections with people in other parts of the world. Through Dr. Kelly's eyes and mind, we see how the Capacha Phase, new to the complex picture of West Mexican archaeology, was born.

Kelly's field work in Colima spanned the years from 1939 to 1973, more than 3 decades of discontinuous study. The body of this report with appendixes, completed in 1975, presents in thorough detail the data recovered during those years of work. An argument can be made for in-depth investigation and immersion in regional archaeology, for without it the Capacha complex might not have been so convincingly recognized and its implications understood. Another fact emerges too. The breakthroughs, the insights that come from investigation, are usually not instantaneous. They result from long hours of plodding and tedious work. The Capacha Phase, though based largely on ceramic evidence and dating from about 1870–1720 B.C., is an example of that, as Dr. Kelly will be the first to admit. The earliness of this complex and the stylistic connections reflected in the pottery with other ceramics from Mesoamerica and northwest South America, single this work out as of fundamental significance to Mesoamerican prehistory. Through these investigations, a window to a new understanding of Mexican-South American connections is opened for us.

Dr. Kelly has long been identified with investigations in Western Mexico, an area stratigically important to the archaeology of the American Southwest, where she has also worked. She has been a Research Associate of the Arizona State Museum for many years, and the University of Arizona's interest in publishing this report in the *Anthropological Papers* was both high and logical.

The text stands intact on the valuable findings through 1975 but Dr. Kelly also has acknowledged later references in her bibliography and end notes. Earlier discussions of the Capacha complex (1972 and 1974) clearly establish her bibliographic priority.

EMIL W. HAURY

ACKNOWLEDGMENTS

My endeavors in the Mexican state of Colima have been scattered sporadically over three decades. The initial field work, in 1939 and 1940, was supported by the University of California at Berkeley. Study of the resulting material indicated that data were not sufficient to establish a convincing chronology, so the collections were packed and delivered to the Instituto Nacional de Antropología e Historia for indefinite storage. Then came years with commitments elsewhere, but eventually I returned to Colima. The 1966 and 1967 field seasons were financed by a grant from the Rockefeller Foundation; those of 1968 through 1971 and 1973, by grants from the National Geographic Society and the Wenner-Gren Foundation for Anthropological Research. To these institutions I am genuinely grateful.

Field work was authorized by the Instituto Nacional de Antropología e Historia. During the 1966 season, Dr. Ignacio Bernal, then director of the Museo Nacional de Antropología, amiably released Prof. Otto Schöndube from his museum obligations for 20 days, so that he might accompany me to the field. Moreover, he arranged for restoration of several key vessels in the Museo laboratory and authorized use of photographs of a number of specimens in the museum.

To several institutions and to many friends and colleagues, I am indebted for courtesies and for helpful information; an effort has been made to acknowledge these individually in the text. Here, however, special mention should be made of the generous cooperation of the staff of the Departamento de Prehistoria and the Departamento de Monumentos Prehispánicos, as well as that of the Sección de Arqueología and of the Bodega de Arqueología, all of the Instituto Nacional de Antropología e Historia.

During 1939 and 1940, Mrs. Marian Cummings provided welcome company and collaboration in the field, as did Prof. Otto Schöndube, in 1966. The summer following the first Colima excavations, Miss Virginia Ross, then a student in anthropology at Yale University, spent some weeks in Mexico, during which time we tried to classify the welter of tomb sherds.

Over the years, I have been indebted to a number of friends in Colima, including the late Don Rodolfo Baumbach; Doña María Ahumada de Gómez, director of the Museo Regional de Colima; Doña Esperanza Padilla Salazar; and Prof. Teófilo Jiménez. In connection with the Capacha material herein reported, acknowledgment is due the several Colima property owners who permitted excavation on their land: Don Jorge and Don Jaime Salazar, Don Urbano Salazar, Don Manuel Amezcua, Don Fidel Valladares, and Don Germán Virgen (Fig. 1, site nos. 2, 3, 6, 8, and 10 respectively); Don Luis Salazar gave limited permission to test his ejido plot (no. 5).

Most of the text figures have been drawn by Doña Catalina Gárate de García; the sketch map is by Don Armando Rodríguez Lara.

Appreciation is expressed to Gail Hershberger for her careful editorial assistance, and special thanks are extended to the University of Arizona Press for their fine production of this volume.

1. INTRODUCTION

In museums and private collections there are hundreds—perhaps even thousands—of archaeological specimens from Colima. Many of them, in the form of hollow animal and human effigies, belong to what I now designate as the Comala phase, which is probably contemporaneous with the early part of the Mesoamerican Classic. The preceding ceramic phase, the Ortices, corresponds temporally to the late Preclassic of Mesoamerica. In the tentative sequence published years ago (Kelly [1944]: 218, Fig. 1; 1948: 65), the earliest phase recognized in Colima was called Ortices. It is now possible to divide this: the earlier part is still designated as Ortices; the later, as Comala (see Chapter 2).

During the past few field seasons, it has been possible to recognize a still earlier ceramic assemblage, contemporaneous with the Mesoamerican Early Preclassic. I have named it Capacha after the ex-hacienda from which came the first material to permit its recognition as an entity. The present paper is designed primarily to put the Capacha phase on record; two preliminary notes have been published previously (Kelly 1972; 1974).

In this paper there will be frequent reference to Preclassic, Classic, and Postclassic, terms commonly used in connection with cultural sequence in Mesoamerica, although not always with precisely the same chronological meaning. I follow Millon (1973: Fig. 12), who designates the span from A.D. 300 to 900 as Classic. Strictly speaking, these handy terms are not really applicable to Colima, whose early affiliations, at least, seem to be less Mesoamerican than South American (see Chapter 4: *Northwest South America*).

The ceramic types mentioned herein are arbitrary groupings—unabashedly subjective—and include material that seems to me to belong together stylistically. In recognizing these types, I tend to rely on form and decoration rather than on paste, inasmuch as the latter may vary with the raw material available in different localities. The term "phase" refers to an identifiable cluster of culture traits—of necessity largely ceramic—with temporal significance.

In ceramic description, I have tried to avoid the ambiguous term "jar" and have followed a useful distinction made by the *campesinos* of Colima. To them, a *cántaro* is a vessel with restricted aperture, proportionately deeper than it is wide, and serviceable only for liquids. In contrast, an *olla* has the mouth of greater diameter and thus is more versatile; for example, it might be suitable for cooking beans. Other terms related to vessel shapes of the Capacha phase are enumerated in Appendix II, *Form*.

At this point, a few words should be said concerning several factors that complicate archaeological investigation in Colima. (1) The ceramics show enormous regional diversity, and (2) a span of 3,000 years prior to the Spanish Conquest is represented. (3) Occupational debris is unaccountably scarce and extremely little has been located for the early phases. (4) Under the circumstances, there is scant possibility of working in stratified rubbish or (5) of obtaining charcoal samples associated with trash.

(6) The association of artifacts found in tombs is not very reliable. Successive use of a given chamber during two, or even three, ceramic phases is demonstrable (Kelly 1978). In part for this reason, I have not made an exhaustive search for untouched tombs. A more compelling reason is that the rural population feels tomb contents its major patrimony. According to local custom, anyone who undertakes tomb excavation pays his workmen the normal wage and, in addition, is expected to give them half the current monetary value of any specimens that result. Sale of large, hollow effigies from a well-stocked tomb may produce 40,000 or 50,000 pesos, half of which would exceed my annual field budget.

(7) Finally, the brisk demand on the part of dealers and collectors, combined with the chronic poverty of the rural population, have converted the latter into aggressive *moneros,* who rely on illicit digging for much-needed supplementary income. (A monero is one who looks for *monos,* figurines, hence one who rifles tombs and burials.)[1] Usually moneros work in teams and dig with appalling zest during the off-agricultural season, and their efforts are both effective and destructive. The entire state of Colima is perforated with their test pits—in some places, as in the cemetery of West Chanal, they are barely 50 cm apart. Many sites are so thoroughly churned that one cannot find an undisturbed area in which to place a test one meter square. However, the end is at hand; unexcavated tombs are increasingly hard to find, and a few discouraged moneros have abandoned the "profession."

Under the circumstances, an investigator must rely heavily on nonsalable discards left by moneros at the scene of rifled tombs; on pottery vessels, which are of little interest to dealers and thus can be purchased at a very modest price; on burials that have been overlooked; on small tests if one can find a spot not thoroughly "worked" by moneros; on surface collections; and on informants' statements. When one cooperative monero remarks casually that he must have "excavated" close to 400 tombs, I am ready to admit that he speaks with far greater authority than I on many aspects of the subject. In several instances, such moneros have been utilized as if they were ethnographic informants, an approach helpful in salvaging a good many data. However, at best, any chronology for the Colima area is bound to be a thing of shreds and patches.

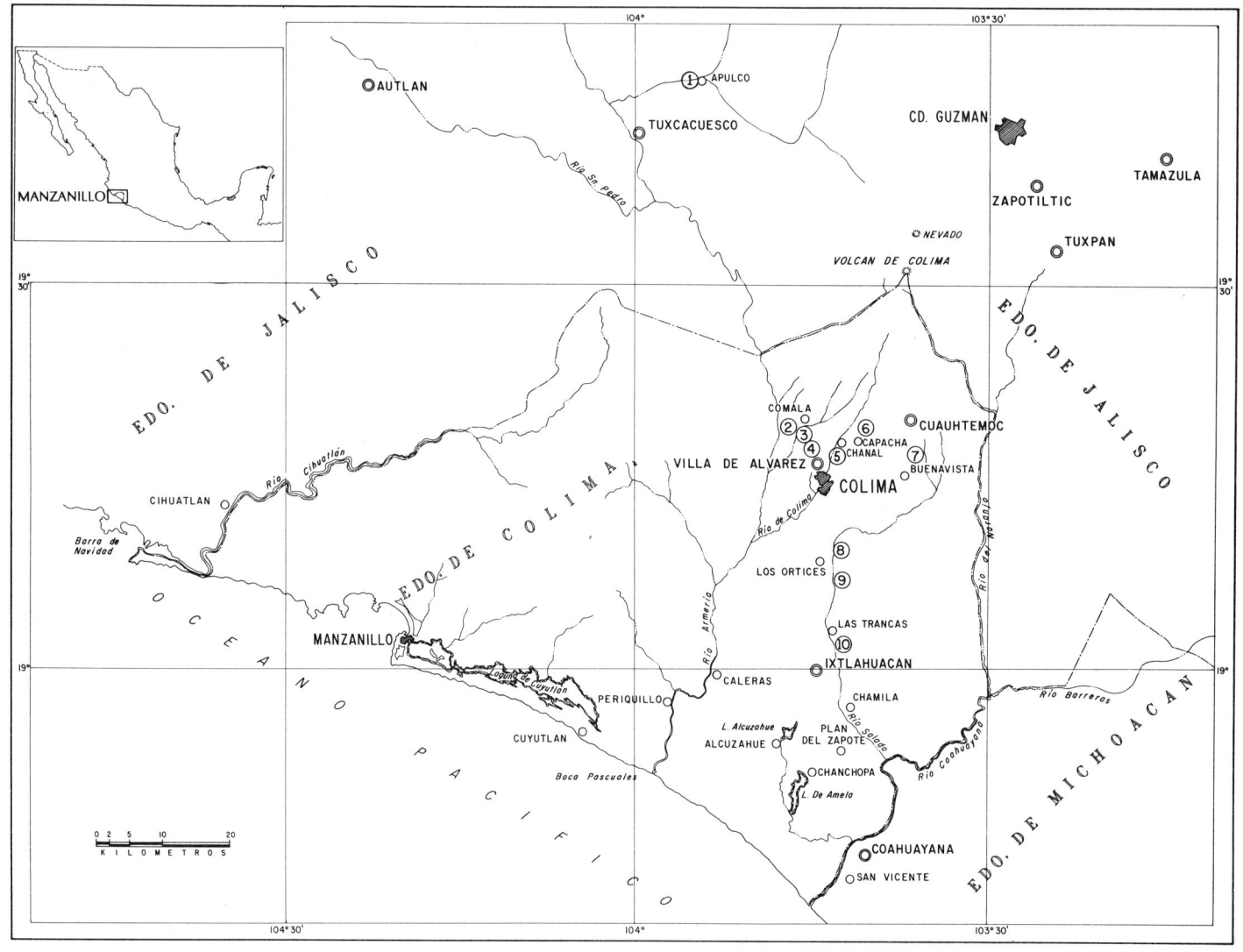

Fig. 1. Map of known Capacha-phase cemetery sites. See pp. 3, 19, 22, 56, 91.

1. Arroyo de San Antonio, Apulco, Tuxcacuesco, Jalisco (Kelly 1949: 83–84, 210). A second Capacha cemetery in the vicinity of Apulco has also been reported (Meighan and Greengo 1974; Greengo and Meighan 1976). (See pp. 18, 22, 39, 48.)

2. La Cañada, Comala. This and the succeeding sites are all in the state of Colima, with nos. *5, 6, 8,* and *9* within the municipal unit of that name. (See pp. ix, 24, 29, 39.)

3. La Parranda (Site A), Comala. (See pp. ix, 24, 43, 61.)

4. Terreno de Jesús Gutiérrez, Villa de Alvarez. (See pp. 28, 44.)

5. Parcela de Luis Salazar, El Chanal, Colima. (See pp. ix, 24, 46, 60.)

6. La Capacha, Colima. (See pp. ix, 18, 46.)

7. El Barrigón, Buenavista, Cuauhtemoc. (See pp. 18, 47, 48, 60.)

8. Terreno de Fidel Valladares, Acatitán, Colima. (See pp. ix, 29, 49, 61, 86, 87.)

9. Las Borregas, Colima. (See pp. 23, 50.)

10. Quintero, Ixtlahuacán. (See pp. ix, 19, 24, 51, 61, 86, 87.)

In addition, a lone specimen of Capacha pottery (Fig. 24 *c*), without precise provenience, is attributed to Autlán, Jalisco (p. 22), and there is likelihood that Capacha ceramics extended far to the northwest, to the lower San Lorenzo valley of central Sinaloa (p. 22).

[2]

2. CERAMIC SEQUENCE IN COLIMA

Before presenting data relative to the Capacha phase, the overall ceramic sequence in Colima should be inspected anew. Many years ago, I published a tentative series (Kelly [1944]: 218, Fig. 1; 1948: 65–66). At that time, the sequence ran, from early to late: Ortices, Colima-Armería, Periquillo. Needless to say, this scheme is very much in need of revision, but so many loose ends are still dangling that I have hesitated to publish.[2]

Colima is a small area of marked regional specialization in pottery, figurines, and other artifacts. Generalization is difficult, for regional boundaries shift from one phase to another. Once the large sherd collections on hand are fully tabulated, it should be possible to indicate graphically the territorial extension of each phase. Until then, statements are impressionistic and subject to revision.

The great Armería drainage system heads in Jalisco, and its upper and middle stretches lie within that state. There its several tributaries unite to form the Río Armería, which enters Colima on the north and follows a north-south course to the sea (Fig. 1), forming what, for want of a better term, is called here the Armería axis. The latter, essentially a north-south strip through central Colima, more or less coincides with the heart of the central Colima archaeological area. While part of the Salado drainage about Acatitán and Los Ortices ties with the Armería axis archaeologically, the vicinity of Ixtlahuacán is of mixed affiliation, and from that point downstream, sites along the Salado relate primarily to east Colima. However, the lower Coahuayana valley aligns less with the latter area than with the Armería axis, at least during the Ortices and Chanal phases.

East Colima presents a complex situation. Its ceramic phases are not firmly defined, much less placed in temporal sequence. Local differences are pronounced and phases are quite distinct from those of the Armería axis. Moreover, the coastal strip of Colima seems, during certain phases, a thing apart. As for the rugged western part of the state, almost nothing is known of its archaeology.

In summary, the following sketch applies primarily to central Colima, that is, to the Armería axis, as defined above, but the discussion will spill over into the eastern and coastal areas when such extension seems warranted. The situation as outlined below probably is generally reliable, but refinement and perhaps some subdivision may come with further study.

PHASES

Capacha

Capacha is the newly recognized ceramic phase to be treated later in the present paper. Its ceramics are the earliest known so far from Colima, and a radiocarbon date based directly on sherds is 1450 B.C., adjusted to 1870–1720 B.C. (Table 1, no. 14). Resemblances to the Opeño phase and to the ill-defined "Tlatilco style" are evident; so also are generalized similarities to early material from northwest South America (see Chapter 4).

Ortices

By no means is Ortices the same today as the phase of that name set up 30 years ago. Named for Los Ortices, the ejido settlement in whose vicinity the diagnostic cluster of wares first was recognized, the phase at that time included what I now consider Ortices, as well as material which in the present revised scheme constitutes a separate, later phase, called Comala. In recent seasons, remains of both phases have been found unmixed.

Evidence of the Ortices phase is comparatively widespread and occurs in concentrated form even in the lower Coahuayana valley. Almost nothing is known of settlements during Ortices times, but shaft tombs definitely are attributable to this phase, in some cases with clear evidence of re-use during the Comala phase (Kelly 1978).

Ceramics

Ortices is dominated by a cluster of closely related ceramic material, usually with moderate polish. A delicate cream to gray "shadow striping" or "wiping" overlies the natural base-colored background, which latter is often smoked, perhaps deliberately, so as to appear dark gray. The shadow-striping technique apparently involves application of a wash which, thereafter, is partially wiped off, and some approximation to a multiple brush may have been used for the removal. In addition to Ortices shadow striped and its variants, another wiped ware (Amoles wiped)[3] occurs in Ortices association and survives in the succeeding Comala phase. The comparatively early occurrence of this decorative technique in Colima may prove to relate to the "blanco levantado" which, somewhat later, cut a broad swath across Mesoamerica (Kelly and Torres 1966).

TABLE 1[1]

Radiocarbon Dates for Various Colima Phases

No.	Ceramic Phase	Laboratory Identification Number	Material	C–14 Age (B.P.)	C–14 Age (A.D./B.C.)	Adjustment to 5730-Year Half Life	Adjustment to Astronomical Time	Provenience
1.	Periquillo	UCLA–1095 C	Charcoal	325 ± 60	A.D. 1600 or A.D. 1450[2]	A.D. 1615	A.D. 1520 (A.D. 1440–1630)	El Columpio, test 2:3 (30–45 cm below surface)
2.	Chanal	M–2338 (Crane and Griffin 1972: 184)	Charcoal	660 ± 100	A.D. 1290	A.D. 1270	A.D. 1260–1290 (A.D. 1200–1380)	El Chanal east, test 3:5 (80–100 cm below surface)
3.	Chanal	M–2334 (Crane and Griffin 1972: 184)	Charcoal	490 ± 100	A.D. 1460	A.D. 1445	A.D. 1410 (A.D. 1340–1450)	El Chanal west, Potrero de los edificios [chicos], test 1, general digging
4.	Armería	M–2339 (Crane and Griffin 1972: 184)	Shell	1260 ± 130	A.D. 690[3]	A.D. 652	A.D. 690 (A.D. 590–880)	El Bajadero, Las Borregas, Burial 11
5.	Colima	UCLA–1095 A	Charcoal	1440 ± 60	A.D. 510	A.D. 467	A.D. 557 (A.D. 460–610)	Potrero del Rancho Nuevo no. 3, Los Amoles; test, cut 5 (90–100 cm below surface) on sterile sand
6.	Colima(?)	UCLA–1651	Human bone	<500 years[4]				La Parranda (site A), Burial 1
7.	Comala(?)[5]	M–2249 (Crane and Griffin 1972: 183)	Shell	2350 ± 140	400 B.C.	470 B.C.	460–440 B.C. (765–395 B.C.)	Purchased, rifled tomb, Rancho del Escritorio, Tuxpan (Jalisco)
8.	Comala(?), Ortices(?)[6]	UCLA–1095 B	Charcoal	1450±60	A.D. 500	A.D. 456	A.D. 545 (A.D. 460–600)	La Loma, Los Amoles, "horno," 40–60 cm below surface, on sterile sand
9.	Comala(?)[7]	UCLA–1627	Shell	1360 ± 80	A.D. 590	A.D. 550	A.D. 620 (A.D. 545–690)	Purchased, rifled tomb, Tierra de Solórzano, Los Ortices. Association allegedly Amoles wiped ware
10.	Ortices(?)[8]	UCLA–1066[8]	Shell	2180 ± 80	230 B.C.	295 B.C.	290–270 B.C. (420–140 B.C.)	Chanchopa, site 2; dump from sacked tomb
11.	Ortices[9]	M–2396 (Crane and Griffin 1972: 185)	Charcoal	2110 ± 140	160 B.C.	223 B.C.	200–170 B.C. (410 B.C.–A.D. 50)	La Paranera, San Vicente, Coahuayana valley (Michoacán), east test, 340 cm below surface
12.	Ortices[10]	M–2341 A (Crane and Griffin 1972: 184)	Shell	1690 ± 140	A.D. 260	A.D. 211	A.D. 270 (A.D. 135–440)	Loma del Volantín, Alcuzahue, rifled tomb
13.	Ortices[10]	M–2341 B (Crane and Griffin 1972: 184–185)	Shell	2330 ± 140	280 B.C.	347 B.C.	410 B.C. (480–140 B.C.)	Same rifled tomb as preceding entry
14.	Capacha	GX–1784	Sherds	3400 ± 200	1450 B.C.	1552 B.C.	1870–1720 B.C. (2110–1520 B.C.)	Terreno de Jesús Gutiérrez, Villa de Alvarez; surface, monero discards from sacked burials
15.	Capacha	UCR–129; inter-lab check UCLA–1888	Charcoal	180 ± 100	A.D. 1770	A.D. 1765	A.D. 1625 (A.D. 1520–1880)	Quintero; beneath "horno" excavated by moneros; obviously contaminated

1. This table lists all radiocarbon dates now available for Colima, except two which refer to the eastern part of the state. There ceramic sequence is not satisfactorily established and so the area is not included in the summary given here. For seven dates, I am indebted to the radiocarbon laboratory of the University of Michigan; for six, to the Institute of Geophysics and Planetary Physics of the University of California at Los Angeles. The remaining dates were obtained commercially. For the several adjustments and corrections, I am grateful to Ing. Joaquín García Bárcena, of the Departamento de Prehistoria, of the Instituto Nacional de Antropología e Historia. We decided not to use the correction for "upwelling" suggested for Pacific-coast shell (Taylor 1970), which would have made nos. 4, 7, 9, 10, 12, and 13 somewhat more recent. Radiocarbon dates were converted to 5730-year half life and astronomical time calculated according to the conversion indicated by Ralph, Michael, and Han (1973). Except for two samples (nos. 7, 11), all come from localities within the present state of Colima.

2. Date said to be susceptible of two readings.

3. The present date is considerably earlier than the A.D. 900–1000 that I had guessed for Armería (Anónimo 1968: 51, foto 41).

4. Obviously, too recent. Associated ceramics suggest the Colima phase, and it was hoped this specimen might date the latter.

5. The sample comes from a site outside the Armería drainage and within Jalisco, but near the Colima border. The excavator's statements concerning the ceramics, plus one vessel still in his possession, led me to suppose that the Comala phase was indicated. However, the resulting date seems rather too early. Under the circumstances, I talked anew with the informant, and it turned out that the shell on which the date is based came from one of two chambers that shared a single shaft. Some of the associated pottery certainly was Comala; some cannot be placed from description. The shell accompanied two hollow figurines which, likewise from description, cannot be assigned to phase. In short, the date in question might apply to Comala but possibly refers to an earlier use of the tomb.

6. Phase affiliation uncertain. Associated sherds are Amoles wiped, a long-lived ware whose association extends from Ortices into Comala. It does not occur in more recent ceramic company, although the radiocarbon date is about what one would expect for the Colima phase. In two tests at an essentially Comala-phase deposit, sherds of

Amoles wiped were infrequent but were scattered almost throughout the pits.

7. This date, based on shell, seems too recent for Amoles wiped.

8. Concerning this date there is long-standing confusion. It is based on a shell fragment from a monero dump adjacent to a presumably looted tomb at Cerro de Chanchopa, site 2. It does *not* belong to the rifled tomb which produced a restorable vessel of Thin orange (Bell 1971: 742), nor can it be associated with the shaky assemblage of east-Colima wares tentatively called Chanchopa. I have tried to check the sherds associated, but, to my mortification, the card that presumably controls location of material in storage is incomplete. My recollection is that this shell fragment was found in association with a large sherd of the local manifestation of Ortices, together with another sherd, unclassified.

The resulting radiocarbon date has been published repeatedly: three times as A.D. 10 (Berger and Libby 1967: 482; Long and Taylor 1966: Fig. 1; Taylor 1970: Table 1, citing Berger and Libby 1967). This version of the date has –240 years applied as correction for "upwelling" (Berger and Libby 1967: 482). The date twice has appeared as A.D. 100 (Taylor, Berger, Meighan, and Nicholson 1969: 18, 27; Kan, Meighan, and Nicholson 1970: 16), with no reason indicated for the change.

9. For the Ortices phase within the Armería axis, there is no date. However, from southeast Colima, there are three certain specimens (nos. 11–13), plus a fourth (no. 10) that is queried. These may be used in the absence of other information. Probably no. 11 is the most reliable.

10. Nos. 12 and 13 are dates based on shell-bracelet fragments from within the same looted tomb. Sherd material left by moneros within the same chamber includes two near Ortices vessels, apparently once grave furniture, and it is these which no. 13 may date. There are smaller fragments of unclassified Red wares, one of which comes from a carinated bowl with red exterior, black interior. There is some chance that this specimen may turn out to be contemporaneous with the Comala phase of the Armería axis. Furthermore, the form of one rim sherd rather suggests the unplaced Manchón wares (Kelly 1978). It will be noted that nos. 12 and 13 represent a time span of half a millennium. Although reuse of tombs can be demonstrated (Kelly 1978), 500 years seems excessive.

Variants of Ortices shadow striped may have a rose-red rim; a simple, geometric design in rose-red paint (Kelly 1978: Figs. 10, 11); or similar motifs in rose and purple-black paint (Kelly 1949: Pl. 19 *c–f*; Kelly 1978: Figs. 12, 13). Shapes include simple bowls with round bottom, some with flaring rim; a few bowls with incurved sides; and larger, open-mouthed vessels, quite deep, flaring at the rim. Cántaros are infrequent, and there are no neckless jars, bottles, or boot-shaped forms. Presence of the "cover" or "tapadera" (see this chapter: *Colima: Ceramics*) is unlikely.

Not shadow striped, but associated with the Ortices phase, is a rose-red ware of somewhat purplish cast, which appears more often as a cántaro than as a bowl. The cántaro may have simple, vertical stripes (Kelly 1978: Fig. 14) or other uncomplicated geometric ornament in purple-black on the upper part of the body; at least one rose vessel with black stripes has, moreover, modeled ornament (Kelly 1978: Fig. 15). There is no clear association of vessel supports with Ortices wares; small tripod bowls, not firmly placed, may be Ortices but more likely belong to Comala, the next succeeding phase.

Small, solid figurines (Kelly 1949: Fig. 80 *d–i*), sometimes depicting action, are Ortices products and continue,

perhaps with greater strength, into the Comala phase. Large hollow effigies, human and animal, are exceedingly rare, but in two private collections are a few such specimens painted in unmistakable Ortices style.

Dating

Table 1 gives three dates for Ortices (nos. 11–13), as well as a fourth (no. 10) that is queried. None comes from the Armería axis and all refer to southeast Colima, but contemporaneity probably may be assumed.

The confusion with respect to no. 10 (Table 1, UCLA-1066) is long standing and is discussed in note 8 of the table. Doubtless the most reliable Ortices date is that from La Paranera, San Vicente, Michoacán, in the lower Coahuayana valley (Table 1, no. 11). It is based on a generous sample of charcoal found 3.4 m below the surface, in a deposit whose overburden has Ortices red on cream as its chief diagnostic ware. This statement is subject to modification, for this large test is not yet fully tabulated. However, the resulting radiocarbon date of 160 B.C. appears reasonable.

The two remaining radiocarbon dates for Ortices (Table 1, nos. 12, 13) are based on shell bracelets found in a looted tomb at the north end of the Laguna Alcuzahue.

The dates are disconcerting: one is A.D. 260; the other, 280 B.C. Although there is ample evidence of sequential use of Colima tombs, a difference of half a millennium in material from the same chamber seems excessive. More than one ceramic complex may be involved, but the little identifiable pottery left in the tomb by the looters is Ortices.

Until further data are available, it may be said that the Ortices phase falls in the centuries just prior to and just following the beginning of the Christian era,[4] although, eventually, it may have to be pushed back in time. However, it probably is safe to regard Ortices as within the temporal bounds of the late Preclassic of Mesoamerica. Ortices is considerably more recent than Capacha, and is unlikely to have followed the latter directly.

External Relationships

Ortices ceramics occur chiefly in the Armería axis and in the lower Coahuayana valley. Only one sherd, considered trade ware, is reported from the Morett site, on the coast, at the Jaliscan border (Meighan 1972: 52). Although there are clear ties between Colima and the Tuxcacuesco area of adjacent Jalisco, in the latter area Ortices wares occur only as trade items (Kelly 1949: Table 24). Nor do Ortices ceramics extend into the Colima hinterland, about Tamazula and Tuxpan (both, Jalisco); Schöndube (1973–74 I: 86) mentions only one possible sherd of "Ortices bichrome." Figurines closely resembling the Tuxcacuesco-Ortices-Comala style seem more widely spread than are the pottery wares that accompany them in Colima and in the Tuxcacuesco-Zapotitlán area.

Ortices appears untouched by influences from the late Preclassic site of Chupícuaro (Guanajuato), although a couple of ocarinas and a flute in one Chupícuaro collection (Natalie Wood 1969: 76, nos. 484–86) are indistinguishable from Ortices-Comala products. That particular collection was formed by purchase, and the specimens in question may actually have come from Colima.

Central Mexico is even farther removed, and Ortices has no apparent ties there. However, two vessel forms reminiscent of the Mesoamerican Preclassic occur in east Colima and may be in part contemporaneous with Ortices (see *Comala: External Relationships,* below).

Comala

With the Comala phase comes a major change in ceramics, although one kind of shadow striped ware (Amoles wiped)[3] continues unbroken from Ortices into Comala. Similarly, the well-known, small, solid figurines, which apparently start in Ortices, are prominent in Comala; reappraisal is necessary, now that it has been possible to separate the two phases.

Although Comala is the era par excellence of the shaft tombs of Colima, it can be demonstrated that some chambers that produce Comala furniture had prior use in Ortices times (Kelly 1978). Practically nothing is known of Comala habitation sites.

Ceramics

The predominant red is no longer a soft rose, but a deep, strong red. Open vessels decline in popularity, and the characteristic form is a cántaro, often elegantly shaped

(Kan, Meighan, and Nicholson 1970: nos. 180, 182, 190, 191, for example), sometimes with effigy tripod feet (Messmacher 1966: láms. 72, 74; Kan, Meighan, and Nicholson 1970: nos. 188, 189: Eisleb 1971: nos. 167–171). Modeling is the chief decoration, usually executed with consummate skill. There also is a black-resist variant of Comala red ware, often with negative polka dots (a poor specimen in Eisleb 1971: no. 244); less frequently there are linear motifs, stylized birds, or even a cartoonlike human face. Shadow striping, so characteristic of the Ortices cluster of wares, disappears except for Amoles wiped. Monochrome engraved wares, primarily funerary, seem to be Comala products; strangely enough, they scarcely appear in the literature (Kelly 1978: Figs. 19, 20).

Many of the large, hollow effigies—animal and human, red or brown—which are so popular among collectors, are Comala products[5] (Kan, Meighan, and Nicholson 1970: nos. 106, 107, 118, 134, 153, 154; Bell 1971: Fig. 31). Regional specialization seems to be considerable but probably never will be known, because practically all the tomb material has been removed by looters, usually to be sold the very day it is excavated. In order to keep a newly discovered cemetery for themselves, moneros cheerfully attribute the specimens they offer for sale to a quite different area.

In addition to the small, solid figures that carry over from Ortices times, other solid figurines—especially some with bowed legs and elongated head—almost certainly are Comala (Alsberg and Petschek 1968: Fig. 48, perhaps correctly attributed to Jalisco, but erroneously described as hollow; Bray 1970: Pl. 12, no. 73; Easby and Scott 1970: Fig. 98; Kan, Meighan, and Nicholson 1970: no. 144; Eisleb 1971: nos. 81–83).

Dating

There is no carbon date for Comala. I tried unsuccessfully to fill the gap with one based on purchased shell (Table 1, no. 7, note 5) from a nearby Jaliscan site. The resulting date of 400 B.C. seems too early and may refer to a pre-Comala use of the chamber.[6]

Some collateral evidence simply aggravates the problem. As noted above, a distinctive shadow-striped ware, Amoles wiped, occurs in both Ortices and Comala association.[3] Occasional sherds have typical Amoles decoration on one face, Ortices on the other; and from two tests in an essentially Comala deposit, fragments of Amoles wiped were infrequent but consistently present almost throughout the several levels. From time to time, Amoles wiped occurs unaccompanied by other identifiable wares, and there are two radiocarbon dates for such isolated occurrences. One (Table 1, no. 9) is based on purchased shell, allegedly from a rifled tomb at Tierra de Solórzano, at Los Ortices. The date is A.D. 590. The other (Table 1, no. 8) comes from charcoal from a shallow "oven" *(horno)* at the site of La Loma, near Los Amoles. It, too, is essentially late: A.D. 500. Such recent dates presumably apply to the terminal era of Amoles wiped and suggest that the Comala phase lasted well into the Mesoamerican Classic. Except in a very few mixed lots, Amoles wiped does not occur in association with post-Comala pottery. A terminal date of A.D. 500 for Comala seems supported by the one radiocarbon date for the Colima phase, which is next in the sequence; it is A.D. 510 (Table 1, no. 5).

External Relationships

There is little ceramic evidence of external ties during Comala times, although one vessel of unknown antecedents, which might be considered a variant of Comala red, has outright Teotihuacan decoration in post-firing pigment (Matos and Kelly 1974). No trade specimens of Mesoamerican-Classic wares have been reported in Comala association, although a few apparently Classic specimens have cropped up in nearby Sayula, in Jalisco (Kidder, Jennings, and Shook 1946: Fig. 195 c; Cerámica Jalisciense 1964: [Fig. 20]), there without association. One explanation may be that practically all Comala material has been extracted by looters and the specimens sold and scattered. Another may be that the Comala phase was so vigorous and so prolific that its potters, engrossed in their own handsome Comala red, were unreceptive to outside influences. The several Teotihuacan elements reported from Colima (McBride and Delgado n.d.) are, by and large, late within Teotihuacan and probably are post-Comala (see *Colima: External Relationships,* below).

In spite of the vast amount of trade pottery found at the Teotihuacan site, nothing attributable to the Comala phase—or to Colima at any time level—has been recognized (personal communication from Dr. Evelyn Rattray, 1974). Moreover, it is significant that the standard vessel forms of the Mesoamerican Classic are scarcely found in the Comala assemblage. One possible exception is a brown cylindrical vessel, notable for its flat bottom and flat cover (von Winning 1969: no. 108, shown without cover). In relief, it has a figure, usually human, within an arch, each end of which terminates in the head of a serpent.[7] Similar specimens, in rose-red ware, are known from east Colima. One other Comala form has vague resemblance to some Teotihuacan products. This is a fairly deep, basically cylindrical vessel, with concave walls. The rim flares; the base is gently rounded and is supported on four solid feet. Exterior walls usually are decorated with engraved diamonds or triangles filled with cross hatch. Otherwise, the Comala phase seems free of ceramic resemblances to the central highlands.

In contrast, east Colima shows considerable evidence of outside contact. Although phases there are not yet ordered, at some time during the shaft-tomb span, if not specifically during the local equivalent of the Comala phase, two east Colima vessel forms suggest derivation from central Mexico. One is the carinated (composite-silhouette) bowl (cf. Early and Middle Zacatenco, Vaillant 1930: Pls. I c; IV d, i, o; VI h, k) that is well represented in surface material collected about rifled tombs. This shape occurs in various wares and apparently runs through more than one ceramic phase. Another comparatively frequent form associated with rifled tombs in east Colima is an olla or a cántaro with four small, vertically placed rim handles (cf. Late Zacatenco, Vaillant 1930: Pl. VII a–c). Rare in the Armería axis, rim handles occur in the Tuxcacuesco area (Kelly 1949: Figs. 68 e, 70 a, b, g, i, j). Theoretically, both the carinated bowl and vertical handles at the rim could derive from northwest South

America (Ford 1969: 101–3, chart 13; Meggers, Evans, and Estrada 1965: Pls. 72 d–g, 112 a; Fig. 54, nos. 8, 9), but the central Mexican occurrences noted above are closer to Colima, spatially and temporally.

Moreover, in east Colima—at a time level supposedly more or less that of the Comala phase—several locally made specimens suggest specific inspiration from central Mexican forms, and at least one important, outright trade piece is recorded. The latter is the large Thin orange vessel reported long ago (Kelly 1949: 195) as a discard from a looted tomb. No longer can this piece be considered evidence for trade relationships with central Mexico in either Ortices or Comala times. The tomb from which it came is near Chanchopa, in southeast Colima, and it contained miscellaneous ceramic offerings, a number of which are still unplaced. When the context in east Colima is better established, the Thin orange specimen should be valuable for cross dating. It now is known that, at Teotihuacan, Thin orange vessels with incised ornament combined with punctation, like that of the Chanchopa specimen, do not appear prior to A.D. 450 (personal communication from Dr. Evelyn Rattray, 1974).

The apparent weakness of Teotihuacan influences in the Comala phase is puzzling, inasmuch as evidence of contact with the Mesoamerican Classic has been reported elsewhere in west Mexico. From Ahualulco, in Nayarit, Mountjoy and Weigand (1975: 353) have sherds of Thin orange, said to tie with Teotihuacan II–III, and Weigand (1974: 127, 130) suggests that "around A.D. 350" the Ahualulco area "appears to have been incorporated indirectly into the Teotihuacán sphere . . . for reasons similar to those for which the Teotihuacán organizations were exploiting the Chalchihuites zone," that is, for acquisition of "rare resources." Perhaps from the latter point of view Colima was not attractive.

As for the northwest frontier, Kelley and Abbott (1966: 325) believe that the "Mesoamerican frontier in Zacatecas and Durango moved slowly northward from about A.D. 100 to perhaps A.D. 900," and more recently Kelley (1974: 20) speaks of a "Mesoamerican expansion, which certainly began in the late Pre-Classic if not earlier." The Alta Vista phase is viewed as "Classic (Teotihuacán IIII [sic]) in age and cultural affiliations" (Kelley and Abbott 1966: 335), although it is recognized that this equation, which involves "pseudo cloisonné," may be disputed. It is pertinent that for Guanajuato, Braniff (1974: 43) reports ceramics typologically Toltec, but chronologically Classic, and that Mazapanoid figurines may appear in west Mexico earlier than might be expected (see *Armería: External Relationships,* below). Such temporal discrepancies would accord with a Toltec origin in the west or northwest, as Kirchhoff (1961) has suggested. Although occupation of the large, well-known site of Ixtepete, near Guadalajara, runs from late Classic into early Postclassic, it is not likely to qualify as a great Toltec hearth (personal communication from Profa. Marcia Castro Leal de Ochoa).

For "Ixtepete and several other large settlements of highland Jalisco, Amapa and several sites in Nayarit, and Chametla in Sinaloa," Meighan (1974: 1258) now sees the Mesoamerican Classic of central Mexico represented in "real force," dating from "some time in the 6th century A.D." He mentions, as evidence, the mound and plaza type of settlement, limited stone architecture and temple mounds, figurines, ear spools, spindle whorls, and red on brown pottery. The suggested date and first appearance of some of these traits in Colima more or less coincide with the end of Comala times and the start of the Colima phase. A summary of the latter follows, at the end of which the discussion concerning relationships with central Mexico is resumed.

Colima

In the scheme published years ago (Kelly [1944]), Colima and Armería were recognized as ceramic phases, but it was not possible to separate them clearly and their temporal relationship was uncertain. To some extent, this still is true. There are at least two, possibly more, ceramic clusters, each with markedly different diagnostic wares, but the residue, lacking these, tends to constitute a common denominator. The situation is complicated by regional differences. However, a good deal of material comes from recent field seasons, and once this is fully analyzed, the panorama may be somewhat less foggy.

Doubtless some *moneros* are correct in their claim of having found Colima-phase pottery in tombs. Nevertheless, the shaft tomb of Ortices-Comala times drops out and is replaced by one with a smaller chamber (popularly called *cueva de alcatraz,* or "pelican's cave"). Its floor plan is somewhat triangular or heart shaped, and the wall opposite the shaft entrance comes to a point. I know this reduced tomb almost exclusively from descriptions provided by informants in the Los Ortices area, and most Colima-phase interments are simple, free burials, in extended position. Sometimes they occur in *ceniceros,* which are considerable deposits, presumably artificial, of white volcanic ash, which contain a heavy concentration of sherds, almost as if cyclic destruction were involved.

During the Colima phase, for the first time, there is clear evidence of small, formal plazas, or several contiguous ones, defined by low, artificial mounds.

Ceramics

Change in ceramics is fundamental and marked by diminished skill. Red on orange or cream *cántaros* (Kelly 1978: Fig. 24), often shouldered, are diagnostic. Decoration is geometric and strongly suggests a multiple-brush technique.[8] A few specimens are well made and highly polished; most are mediocre (Eisleb 1971: no. 232; his 231 *a, b* may be a late variant, perhaps Armería rather than Colima). Elaborate modeling disappears, as do tripod feet; a very few low pedestal bases apparently are harbingers of the Armería phase. Resist painting is absent. Shadow striping is abundant, but altered, and the typical Colima-phase specimen so decorated is an olla, occasionally a basin, with red throat; the exterior is matte, covered with cream shadow striping on the natural-brown base (Kelly and Torres 1966: foto 33; Kelly 1978: Fig. 25).

Grinding bowls, or *molcajetes,* are recorded for the first time, although they are said to appear earlier on the Colima-Jalisco coastal strip (Meighan and Foote 1968: 111; cf. Meighan 1972: 36–37). Those typical of the Colima phase are simple bowls, usually somewhat incurved; the base is gently rounded and normally without any support. Color is red or brown to black, and below the exterior rim there is simple, decorative broad-line incision, sometimes combined with punctation (Kelly 1978: Fig. 26). An unslipped circle in the center floor is jabbed to provide a rough surface, presumably used in food preparation. Other gouged-floor molcajetes, without incised ornament, are abundant and may represent a regional or temporal difference within the phase.

For the first time, the "cover" or *tapadera* appears with certainty; it might be earlier and assuredly is of very different style in east Colima. The tapadera of the Colima phase is of unpolished plain ware, usually tetrapod, in animal form (Eisleb 1971: no. 50; Kelly 1978: Fig. 27). A few specimens are painted with simple geometric motifs in white, and there is one case of red paint, zoned by incision. The tapadera is thought to have been placed over burning incense, and some specimens have the underside blackened; a good many do not, perhaps because they were made as grave furniture and were deposited unused with the corpse.

As noted above, it is possible that a low pedestal base first appears during the Colima phase, and the same may be true of a potstand or stool, with a medium-height pedestal support.

The Colima phase is marked by a surprising and apparently total absence of figurines. Spindle whorls appear sparingly.

Dating

The one radiocarbon date for the Colima phase is A.D. 510 (Table 1, no. 5). A sample of human bone, expected to provide a second date (Table 1, no. 6), inexplicably turned out at less than 500 years of age, hence must be disregarded.

External Relationships

A date in the sixth century of our era would make the Colima phase contemporaneous with the Xolalpan (Teotihuacan III–IIIA, Millon 1973: Fig. 12). This coincides roughly with the time Meighan (1974: 1258) postulates a great push westward of central Mexican culture, after which "the predominant West Mexican culture . . . was merely a variant of the strong Central Mexican tradition, and West Mexico was a participant in the culture sphere dominated by Teotihuacan and its successors." Such major impact is not very apparent in the Colima area, at least not until considerably later. There are, to be sure, some of the traits suggested by Meighan: plazas outlined by low artificial mounds, spindle whorls, and red on cream pottery.

Although one typical Colima-phase ware is red on cream or red on orange, specific Teotihuacan resemblance is slight. Exceptions are the wide-mouthed jars illustrated by McBride and Delgado (n.d.: lám. 1, nos. 1–3, 6). The form of these vessels certainly is reminiscent of Teotihuacan, but perhaps on a time level somewhat later than the sixth century; in fact, Bennyhoff (in McBride and Delgado n.d.: 2–3) believes one of the jars may be a Mazapan-Coyotlatelco hybrid, hence possibly post-Metepec and consequently Postclassic. In the Colima series, the several jars are atypical. Tentatively, I should guess they might belong somewhere on the Colima-Armería transition, but it must be repeated that the red on cream pottery current during these times needs a great deal of additional study. In contrast, the censers figured by McBride and Delgado (n.d.: 5–6, lám. 1, nos. 7–8 a) look earlier to Bennyhoff, although without "specific equivalents" at Teotihuacan.

With respect to the other assorted specimens illustrated by McBride and Delgado, it must be noted that (1) quite a number are without data of provenience; (2) in some specific cases the reported provenience conflicts with information given me by the owners of the specimens; (3) all are without ceramic association; and, as the authors say, (4) most look fairly late in Teotihuacan terms. It may be suspected that a number of the pieces are from east Colima or from the coastal strip, where ceramic phases are not entirely parallel to those of the Armería axis.

In summary, there is no indication that Colima was overwhelmed by influences from central Mexico in Comala times and, although such contact may be perceptible during the Colima phase, it can scarcely be considered spectacular. East Colima, in contrast, seems to have been more open to outside influences, and eventually it should be possible to view these in clear perspective. As matters now stand, it appears that any major influences from Teotihuacan that moved westward during the sixth century of our era (Meighan 1974: 1258) were channeled north of Colima. They may have affected parts of Jalisco-Nayarit-Sinaloa directly and the Chalchihuites area less so, but they seem to have reached Colima in attenuated form.

Armería

It already has been noted that the Colima and Armería phases constitute a continuum. Polar differences are obvious but, when the diagnostic traits of each assemblage are lacking, distinctions are uncertain. Assuredly there are internal stylistic and temporal differences, still not documented adequately, and these are complicated by regional variations.

No longer is the shaft tomb utilized. Some burials are direct, extended; others, tightly flexed, are in small subsoil pits, with the walls somewhat undercut. For the first time, there is evidence of major engineering enterprise. Some Armería sites are defensive, and in several cases hill slopes have literally been carved to provide a flat spot for a plaza, either fully surrounded or enclosed on three sides by low mounds.

Ceramics

Apparent derivatives of the Colima red on cream or red on orange cántaro continue, but these need further study, both stylistically and with respect to association. Painted decoration is geometric, sometimes untidily executed, much of it seemingly done with some sort of multiple brush.[8] Vessels identical with Morett black on white (Meighan 1972: Pl. 35) may prove to be among the late forms of Colima red on cream; there is some suggestion that the latter color combination was intended but that the red turned black during firing.

Several diagnostic, closely related ceramics are grouped as Armería cream wares. Paste is comparatively fine and some vessels are quite thin in section. There may be a partial or total cream to orange slip, occasionally with delicate designs in black resist. Not all specimens are slipped; sometimes color is cream and red, or orange and red, applied in bold, usually geometric designs that all but hide the underlying base ware.

Form varies. One characteristic shape is a medium-sized bowl, with rounded base and no marked shoulder, but with walls slightly incurved or sloping gently inward in a straight line. A showy form, called by moneros a *florero* (flower vase), is an open plate atop a fairly tall pedestal base. Technically, most floreros are molcajetes, with coarse punches or longitudinal gouges that roughen a small central portion of the floor. In some cases, an instrument (a small stick with rough tip?) has been dragged intermittently across the floor to produce a series of short striated gashes (Meighan 1972: Pl. 16 a). The molcajete-floor treatment seems in some specimens to be no more than a stylistic formality; in others, there is evidence of wear. Not all Armería collections include floreros, and either a temporal or a regional difference may be involved.

The same is true for two other molcajete types. One, apparently a deteriorated version of the Colima incised molcajete, has broad-line incised ornament, sloppily executed, below the exterior rim. The other is a simple, open bowl, with rounded or nearly flat bottom. The paste is comparatively fine; interior walls are red, sometimes with quite elaborate resist ornament (Eisleb 1971: no. 238), and the center floor is finely punched or gouged. Such specimens may belong to a regional variation of the Armería phase, but more study is necessary.

Of special ceramic products, the tapadera persists but is poorly made. The potstand or stool, with a medium-tall pedestal base, which sometimes has simple cut-out decoration, appears in Armería association but seems to start in Colima times. Another specialized product that may be Armería is known to moneros as a *piña* (pineapple) or *pitahaya* (probably an *Epiphyllum*) (Kan, Meighan, and Nicholson 1970: 204). It is a deep bowl that rests on a pedestal base, with a rim rising in four peaks. The exterior has spiked or hobnail ornament, and the entire surface is covered with a thick matte cream slip. The one example at hand was collected in fragments from the surface of a looted cemetery; monero information indicates Colima or Armería association, probably the latter.

Figurines occur during the Armería phase but seem limited to sites in the lower Armería drainage. Location and possibly style suggest relationship to figurines from the well-known Playa de los Monos (or Playa del Tesoro) site on Manzanillo bay. The Armería examples from that vicinity are made by hand and are flat and solid, with elaborate, highly characteristic detail in appliqué.

Stone Sculpture

Human and animal figures, in a style that is vigorous but not beautiful, come in secure Armería association (Anónimo 1968: 51–52 and foto 41, which shows specimens from my excavations).

Dating

The lone date for the Armería phase is A.D. 690 (Table 1, no. 4), somewhat earlier than I had guessed (Anónimo 1968: 51). At least, it agrees with the impression that Armería is more recent than its close relative, the Colima phase.

External Relationships

Although the diagnostic Armería ceramics are distinctive and readily recognizable, apparently they do not extend much beyond the limits of modern Colima, and even there they tend to be concentrated near the coast.

The one carbon date for the Armería phase places it within the temporal limits of the Mesoamerican Classic, but my impression is that it may be quite late within that division and probably extends into the early Postclassic. Somewhere at this general time level the west coast—from Nayarit to Sinaloa—was inundated by a spate of Mixteca-Puebla influences. Colima was bypassed, as it had been during Classic times, and I know nothing from there to equate with the handsome pictorial Cholula-like vessels from coastal Nayarit (von Winning 1976) or with other Nayarit-Sinaloan manifestations of this Mesoamerican infusion from central Mexico.

Nevertheless, in Armería association, there is some evidence of ties with the Tula-Mazapan phase of the Postclassic in central Mexico. Three simple, unpolished plates resemble "Mazapan bowls"—not the "wavy-line" variety but that which Linné (1934: 77, Fig. 89) calls his second type. Also in Armería association are two fragments of Mazapanoid figurines, plus a third dubiously classified as such.

In the vicinity of Tuxcacuesco (Jalisco), Mazapanoid figurine fragments belong to the Coralillo phase (Kelly 1949: 119), presumably contemporaneous with Armería. Pottery of the same Coralillo phase includes a characteristic ware called La Loma red on brown (Kelly 1949: Pl. 16 a–d, Fig. 42), whose color scheme and decoration suggest the well-known Coyotlatelco ware, although the molcajete floor is foreign to the latter. At Tula, Coyotlatelco pottery is abundant, appears to be antecedent to Mazapan, and is not Toltec associated (Rattray 1966: 181). This rather suggests that the Coralillo phase may incorporate some immediately pre-Mazapan influences in addition to Mazapan-like figurines.

Further details, apparently confirmatory, come from the Chapala area (of Michoacán and Jalisco). Lister believes that the red on brown pottery of his Chapala complex has significant resemblances to Linné's type two Mazapan ware, while Mazapanoid figures at Cojumatlán run through both the Chapala and Cojumatlán phases (Lister 1949: 37, 60–61); they seem absent in the Tizapan phase, the most recent of the sequence (Meighan and Foote 1968: 123).

Apatzingán (Michoacán) provides an apparently discordant note in that Mazapan-like figurines are allocated to the Chila phase, which is believed, on the strength of an iron blade found deep in a mound of that affiliation, to come to contact times (Kelly 1947: 29). However, there is strong suggestion of a time difference within the Chila phase, and the Mazapanoid figurines are found with Chila polychrome, which seems to represent the earlier strain (Kelly 1947: 42–43).

Little is known of the last few centuries of pre-Conquest times in the vicinity of Etzatlán (Jalisco, near the Nayarit border). The Huistla material from there (Glassow 1967) stops short of the time level in question, although Long (1966: Table 12) shows a Huistla phase, with Mazapanoid figures, as the terminal one of the sequence.[9] From nearby Ixtlán del Río (Nayarit) come several Mazapanoid fragments that Gifford (1950: 233, Pl. 28 *c–f*) assigns to his Middle period.

Fifty-three Mazapan-like figurine fragments are reported from Amapa (Nayarit coast), and the fragment of a mold indicates some local manufacture (Grosscup 1961: 392). It seems significant that no Mazapan-like pottery comes from the Amapa excavations (personal communication from C. W. Meighan, 18 May 1975), and the figurine style, which perhaps had religious connotations, may have a wider distribution than do the pottery wares associated with it at the supposed source—as in the case of the Tuxcacuesco-Ortices-Comala figures previously mentioned. A guess date of A.D. 900 has been assigned the Mazapanoid figures from Amapa (Pendergast 1962a: 370).

Except for the unsubstantiated placement of the Mazapanoid figurines at Etzatlán, it appears that such specimens are widespread in west Mexico and that they tend to occur in the penultimate phase. This agrees quite well with the data from central Mexico. However, it must be remembered that there is no overall typology for Mazapan figurines and their relatives, and that nowhere do we have more than guess dates for their appearance. There is some suggestion that in the west (at Cojumatlán, Michoacán, and in Colima) these dates may be somewhat earlier than the one usually assigned them in central Mexico (see *Comala: External Relationships* for possible implications). Once excavations at Tula have been completed, it is hoped that the resulting information will provide a firmly anchored chronology at the point of possible—but by no means certain—origin.

Figurines of Mazapanoid style are not hard to recognize, but sherds of Mazapan-like pottery might easily be relegated to the inevitable lot of "Red on cream, unclassified." Although wavy-line Mazapan has not yet been reported from the west, specimens reminiscent of Linné's second type come from the Armería, Chapala, and Cojumatlán phases. These occurrences, likewise, seem somewhat earlier than the date usually assigned in central Mexico. As far as the feeble Armería incidence is concerned, it must be remembered that the one date for that phase is based on shell, hence somewhat suspect, and that application of the "upwelling" adjustment would make the date even more recent.

In summary, about all that can be said is that the Armería phase and several of its presumed contemporaries in the west produce occasional Mazapan-like figurine fragments and, less frequently, suggestions of one kind of "Mazapan red on yellow" pottery. If these result from central Mexican influences, as is generally assumed, such influences seem to have been thin and widely spread. As will be seen below, other Tula-Mazapan traits appear in greater strength later (see *Chanal: External Relationships,* below).

Chanal

Next comes an evidently late phase that has only in recent years been added to the Colima sequence. It is associated primarily, although not exclusively, with El Chanal, a site of urban dimensions, which sprawls on both sides of the Río de Colima, about six kilometers upstream from the state capital.

In the whole Colima area, El Chanal is the only locality that suggests an important ceremonial center. It is unique in having had five "pyramids" with carved stone risers (Rosado Ojeda 1948: Fig. X; Castellanos 1952: 35). There is a consistent disposition of mounds in the immediate vicinity of at least two of these structures, and large-scale mapping would doubtless show other regularities. Mounds are abundant—some seem to be largely natural, some are apparently modified by man, and some are totally artificial. Occasionally, in the modern ejido that sits atop East Chanal (that is, the part of the site east of the Río de Colima), there are exposed sections of old cobble masonry, well set in mud, and one mound west of the river has eroded to disclose a tiny bit of red stucco. The proportions of one depression in West Chanal suggest a ball court. The special features mentioned occur only at El Chanal, except for the "ball court" (see Appendix I, no. 7).

The area is attractive scenically and has perennial water but little arable land. Evidently El Chanal was favored for settlement throughout pre-Conquest times. At least, interments of both the Capacha (Fig. 1, no. *5*) and Comala phases have been found on its downstream fringes, and Comala vessels reputedly come from its upstream limits. Of all sites in the Colima area, El Chanal is the one most likely to yield stratigraphy if excavated systematically and intensively—an enterprise that would be both lengthy and costly.

Another outstanding—and unique—aspect of El Chanal is a cemetery west of the river, which produced quite splendid ceramic and other grave furniture. Because of the comparative abundance of gold , the plot has literally been destroyed by frenzied monero activity, stimulated by avid dealer-collector interest. A very few remaining surface sherds, plus occasional specimens in private and museum collections and descriptions from informants, indicate that offerings included variants of Autlán polychrome (Kelly 1949: frontis.), as well as pottery exotic in local context (see below). Artifacts of copper, gold, and silver accompanied the interments, and one monero reported that some corpses had been covered by a thin sheet of gold. The obvious implication is of socio-economic stratification, probably with religious concomitants.

A second cemetery of West Chanal seems also to have associations with religion. In the early 1960s, looters removed from it about 50 "singing" effigies. These, it is said, were accompanied by "warriors" of similar style (Tamayo 1973: Fig. 49) and by heavy pottery braziers, but not by Tlaloc "urns" (see *Ceramics,* immediately below). Copper came from this particular cemetery, but no gold.

Likewise suggestive of religion, or perhaps of war, is a low (1.5 m) mound in West Chanal, composed largely of small calcined fragments of human bone, probably from several hundred persons. Clearly the bodies were cremated elsewhere; thereafter the burned remains were collected and then dumped on this mound. The latter contains fragments of prismatic obsidian blades that have been through the crematory fire, as well as sherds of braziers and Tlaloc "urns" (again, see below), plus fragments apparently of the "singing" figures mentioned above.

Ceramics

Years ago, surface material at El Chanal was scant and difficult to place in the Colima series. Now, owing to monero depredations, sherds are abundant, and several small tests that I made produced modestly. Surface and excavated collections have not been studied closely, although I have done spot checking; comments below are subject to correction.

Except for grave furniture, Chanal pottery from both sides of the river is rather humdrum. Painted wares are primarily tripod bowls, with a noncontinuous flange on the exterior below the rim, or lower. A few specimens are molcajetes, the floor with incised diagonal cross hatch. Tripod supports are usually solid truncated cones; hollow, rattled feet seem infrequent. Painted decoration is chiefly on the interior vessel walls and is red on brown or, rarely, a near-Autlán-polychrome combination—that is, white on red, plus red overpaint on the white, resulting in orange. In a few cases, this distinctive painting is applied over prior engraving or incising. Except for burial offerings, other painted pottery is white on red, limited to half a dozen sherds from East Chanal; these may prove to be significant chronologically (Kelly 1945b: 31, Pl. 2 *i–r*; cf. Meighan and Foote 1968: 94, Fig. 14).

Burial furniture is more decorative. Some is very close to Autlán polychrome, as noted above. Atypical, but related, are deep, incurved vessels, with vertical body constrictions forming lobes; supports are tall, rattled tripods, and the fairly elaborate incised-engraved and painted decoration is exterior (Corona Núñez 1960a: [lám. 23]). From monero descriptions, it appears that some pottery has decoration in post-firing pigment; a "toad of many colors" is mentioned, and a small pale-green fragment comes from the surface of the gold-producing cemetery west of the river. One vessel attributed to the same cemetery seems to be a copy of Plumbate, and a handled censer, two-footed, with cut-out decoration on the bowl, is said to have the same provenience.

Other ceremonial vessels come from monero activity at El Chanal, from both sides of the Río de Colima and not necessarily from cemeteries. First may be mentioned a heavy censer or brazier, quite similar to "Chila" specimens from Apatzingán (Kelly 1947: Pl. 8), except that Chanal examples have large, hollow, slightly bulbous tripod feet instead of hollow cylindrical supports. A second item, presumably ceremonial, is likewise a censer or brazier. It is a large, bottomless cylinder, known locally as an "urn" *(urna).* The exterior walls are covered with elaborate appliqué, frequently in the form of grotesque Tlaloc motifs, with deep eyes almost the size of saucers (Médioni and Pinto 1941: no. 229, classed as Aztec; von Winning 1969: Pl. 121; Bell 1971: Fig. 36, shown inverted; Schöndube 1974: Fig. 10 *a*). A third special Chanal product, possibly not ceremonial, is a canteen-shaped vessel, with body flattened fore and aft; on one of the broad surfaces is modeled an almost life-sized human face.

A few aspects of the unglamorous pottery warrant mention. In some cases, cántaro fragments, both Red and Plain unclassified, are mold made, with a horizontal seam at midbelly (cf. Llano red ware, Kelly 1947: 77). Some cántaro and olla sherds are red at the throat and at the rim, below which is a broad unslipped zone, striated as if the pot were wheel thrown. Heavy, arched handles, circular in section (Kelly 1947: 81, no. 4; 1949: 55, Fig. 38 *b*), may belong to a slant-mouthed dipper (see *Chanal: External Relationships*), although identifiable fragments of such shape are very rare. The usual comal or griddle has a low vertical wall, with a sharp wall-base angle and with interior walls and floor red slipped.

Little is known of Chanal figurines. No Mazapanoid specimens have been reported. Aveleyra (1964: [lám. 118 *a, b*, 123]) illustrates an extraordinary figure with all-over incision and an effigy vessel, both presumably from the gold-producing cemetery of West Chanal. A second burial ground in that same part of the site has produced quite a different kind of figure, apparently limited to two representations: a "singer" and a "warrior." Specimens are comparatively large (one singer, 61.5 cm tall), hollow, of coarse clay. They usually are unpainted, although some have rough decoration in unfired pigment. The arms of the singer extend outward at the sides, with the elbows bent, so that forearms and hands are aloft (Corona Núñez 1960b; Furst 1965a: Fig. 30; Tamayo 1973: Fig. 30; Schöndube 1974: Fig. 8 *a*). It has been suggested frequently that these are female Xipe representations; moreover, both stance and arm position are reminiscent of some of the "smiling" faces of central Veracruz (Heyden 1970). Warrior figures apparently are less common but generally similar in style (Tamayo 1973: Fig. 49).

There are no spindle whorls in the present Chanal collections, and spinning may have been done with a weight of some perishable material. However, I remember vaguely that a few small, globular pottery artifacts—beads or whorls—were offered for sale. Stamps and seals, both flat and cylindrical, are scarce in Colima. I have seen few, and none with the flatiron handle which, in the Field collection, is evidently the Chanal hallmark (Field 1967: 38).

Metal

Metal appears definitely in Chanal context and, likewise, in the Periquillo phase, to be discussed next; at present, there is no authenticated occurrence prior to these times.

Tests at the site of El Chanal provide a few copper specimens in the form of an eyed needle, a small pin or hook, and a couple of bells.[10] From outlying sites with Chanal or near-Chanal affiliations comes a small assortment of bells, acquired by gift or purchase; two of particular interest are of simulated wirework, with what appears to be a Tlaloc face in relief on the resonator.

Gold was abundant in one cemetery of West Chanal, scarce elsewhere. Informants report it was found mostly in the form of small, undecorated sheets, paper thin. However, bosses about the size of a bottle cap seem also to have been common. Some bosses are cut in one piece with a truncated triangle plus a complete isosceles triangle, the latter extending downward from the disk; raised rims bound these geometric components. The most impressive gold items attributed to the West Chanal cemetery are cut from a thin sheet and decorated in repoussé. Small human figures (to moneros, "Aztecas") are mentioned. One private collection has a tiny bird head, perhaps an eagle, but more likely a condor, and a pair of L-shaped ornaments (7.2 cm in height), with detail in repoussé. The general effect of the latter pieces is vaguely Mesoamerican, yet unlike anything I know from elsewhere and without parallel in Pendergast (1962a; 1962b). The same collection includes half a dozen small globular pottery beads covered with gold leaf, as well as two small gold bells of simulated wirework, evidently cast. Silver seems scarce; the very few specimens I have seen are bosses or modified bosses.

Stone

El Chanal is supplied generously with prismatic obsidian blades—most, trapezoidal in section, but some, triangular. It is also the alleged source of the only obsidian core I remember having seen in the entire area.

The stone risers of the "pyramid" staircases have been mentioned above; of the total of 34 stones with "Aztec hieroglyphs," five are said to represent Tlaloc, while the others are "calendar signs" (Castellanos 1952: 35) or unidentified representations. Other sculptured stones—particularly human faces, some of Tlaloc, on roughly rectangular slabs—come from the general area but are not specifically attributable to El Chanal; their association is unclear.

Mano and metate fragments are surprisingly infrequent. At the moment, it is preferable not to comment on them except to note that footed metates are known from El Chanal, and one has on its underside a human figure with headdress in low relief; the style rather suggests Toltec work.

Dating and Distribution

Two radiocarbon samples from El Chanal are essentially late: A.D. 1290 and A.D. 1460 (Table 1, nos. 2, 3) and place Chanal subsequent to the Colima-Armería phases. Moreover, from the two lowest levels of my Test 2 at El Chanal (cuts 8 and 9, 150–200 cm), which were overlain by Chanal-phase pottery, there is a total of 25 sherds of Colima-Armería wares—weathered, but with identification indisputable. This is one of the few instances of clear stratigraphy in the entire Colima area.

What may follow Chanal is problematical, and the question should be considered in connection with the unusual spatial distribution of the phase. There are two main foci: El Chanal proper, and the lower Coahuayana valley, the latter incidence presumed primarily on the basis of burial furniture, for no major occupational zone has been located there.[11] Chanal material is scarcely represented in eastern Colima, in most of the Salado drainage, and in the Armería axis downstream from the Río de Colima.

Surface collections—which still have not been fully studied—suggest two explanations that should be explored. In the first place, a number of sites in the vicinity of El Chanal relate to it generically, although they lack the decorated wares and the characteristic elements presumed to be ritualistic. Their unglamorous sherds, which might be designated as "rustic Chanal," could represent a time difference, or they could belong to comparatively modest satellite communities that were contemporaneous with El Chanal but lacked the latter's more decorative products.

In the second place, in the Armería axis, downstream from El Chanal and upstream from the chief settlement of the Periquillo phase, a few modest sites have surface material that looks "late." Comal fragments are plentiful, as is obsidian, yet the pottery does not relate clearly to either Chanal or Periquillo.

Little is known of Colima archaeology during the last few centuries before the Spanish impact. In the entire area, no site can be identified surely with a pueblo that was a functioning entity at the time of the Spanish aggression. A good many modern settlements appear to occupy their traditional localities and thus may hide the archaeological evidence, but this certainly cannot explain the apparently complete absence of identifiable contact sites.

The more recent of the two radiocarbon dates for El Chanal is A.D. 1460, barely 60 years before the Conquest. Even so, it seems unlikely that El Chanal was a large, operating native center in the sixteenth century, for apparently it is not mentioned by Lebrón (1951). Using the latter source, Sauer (1948: Map 3) identifies native and Spanish settlements of the mid-sixteenth century, and the stretch immediately north of the Spanish Villa de Colima is conspicuously untenanted. By Lebrón's time, this zone presumably had been cleared of native pueblos to permit its use as grazing land (Sauer 1948: 46).

The Villa de Colima that figures in Lebrón is in its second location,[12] traditionally at or near (X)almolonia, the latter placed by the Suma de Visitas (Paso y Troncoso 1905: no. 682, under the name of Molone) half a league from Colima. Modern Villa de Alvarez, identified with (X)almolonia, now merges with the city of Colima on the northwest and thus is comparatively close to El Chanal and even closer to a large site that appears to relate to "rustic Chanal." Populated or not in the early sixteenth century, El Chanal doubtless was one of the places with "fallen buildings and ruined houses" viewed by Alonso Ponce (1873: 109–10) and his party in 1587, as they journeyed from the Villa de Colima to Tonila.

If El Chanal does not reach Conquest days, there is the possibility that "rustic Chanal" might be an epigonal form of the Chanal phase. Or the same might be true of the unimpressive and unplaced "late" sites scattered particularly in the stretch between El Chanal and Periquillo concentrations. Furthermore, the last-named phase, known largely from the lower Armería drainage, is obviously related to the Chanal phase and should be, at least in part, contemporaneous with it. The one radiocarbon date for Periquillo is ambiguous, but close to Conquest times (see *Periquillo: Dating*, below).

Chronological aspects of the Chanal phase will be discussed further in connection with its relationships to other areas.

External Relationships

The Chanal phase casts oblique light on the murky era a few centuries prior to the Spanish Conquest, but perhaps it confounds rather than clarifies. On several scores, it represents a sharp break in local cultural tradition, with the implication of major socio-economic-religious change. For the first and only time in the Colima area, there are masonry "pyramids"—very modest ones, to be sure—in conjunction with an extensive and presumably major religious center. Influences from the outside are implied.

On the score of external ties, there is no indication in Colima—in Chanal or other association—of the cultural assemblage usually associated with the late "Tarascan" occupation in the Michoacán lake area. There are, for example, no mounds of the specialized Tzintzuntzan type, no elegant and sophisticated teapot or stirrup-vessel forms, no polychrome miniatures, and no Tarascan-like pipes.

Chanal ties with central Mexico will be examined below, but first it may be pointed out that the Chanal phase shares a few specific ceramic traits with several phases thought to be roughly contemporary in adjacent areas of Jalisco and Michoacán: the Toliman, at Tuxcacuesco, Jalisco (Kelly 1949); probably the Autlán, at Autlán, Jalisco, despite gaps in the surface collections (Kelly 1945b); the Amacueca, of the Sayula basin, Jalisco (Kelly n.d.); the Chila, at Apatzingán, Michoacán (Kelly 1947); and the three phases at the Chapala shore sites of Cojumatlán, Michoacán (Lister 1949) and Tizapán, Jalisco (Meighan and Foote 1968).

The ceramic elements that seem common to these phases are: the comal, or griddle; the skewed or slant-mouthed dipper;[13] the large, arched handle, associated vessel form uncertain, but sometimes found on the skewed dipper. Evidently these traits are interrelated, but certain others seem to appear in arbitrary association. They are: the horizontal, midbelly seam, indicating use of a mold for some ollas and cántaros; and the open bowl, tripod, often with wall flange, sometimes with incised floor, and with painted decoration concentrated on the interior wall. Bowls of this general description (which include Autlán and Cojumatlán polychrome) occur, with wide local variation, over an enormous area at this general time level. Perhaps the brazier should also be added to the list; it approximates, except in foot form, the specimen illustrated for the Chila phase of Apatzingán (Kelly 1947: Fig. 29). The slant-mouthed dipper and arched handle may be local developments. The midbelly seam probably is not, for it occurs at Teotihuacan (personal communication from Dr. Evelyn Rattray, 1974) and at Tula (observation, two specimens, collection of University of Missouri–Columbia at Tula).

Apparently Chanal shares all these elements, and moreover has several that strongly suggest central Mexican influences. One is the perforated censer with handle and two feet, found also at the Chapala shore sites (Lister 1949: 57, Fig. 23 c, d; Meighan and Foote 1968: 114). In addition, Cojumatlán has produced a small Tlaloc-effigy censer (Lister 1949: Fig. 24) that may relate to the over-sized specimens with Tlaloc representations known from El Chanal and Tula.

Some of these traits suggest Tula-Mazapan influences at El Chanal; others give the same impression, but perhaps spuriously. For example, the comal is common to the Chanal and Tula-Mazapan phases; yet the Chanal comal has a sharply rising wall, with a pronounced angle, as do Coyotlatelco specimens (Rattray 1966: 115, Fig. 1 h, i), while Tula-Mazapan griddles have a low rim, but no real wall (Acosta 1956–57: Fig. 20, no. 1). The tripod bowl, sometimes a molcajete, is another common element, but one so generalized and widespread as to be almost meaningless at present.[14] As for Chila-type braziers, numerous examples are known from Tula and seem to be both Tula-Mazapan and Aztec in association. The same may be true for the perforated censer with two feet and for the great braziers with appliquéd Tlaloc features.

Time allocation of clay stamps is uncertain. Purchased material from the Coahuayana valley—apparently belonging to a time level somewhere within the Preclassic of Mesoamerica—includes four circular stamps of spiral motif, with a stub handle. An anthropomorphic example in a private collection closely resembles a stamp illustrated for Tizapan (Meighan and Foote 1968: Fig. 40 b); it is attributed to El Chanal. Field (1967) reports numerous specimens said to be from the same site. He mentions none in Tula-Mazapan association, but a few fragments come from Tula (information from Dr. Richard Diehl, Dr. Robert Cobean). They may continue into more recent times, for Field (1967: 23) speaks of "Aztec influence" in a collection from the "Post-Classic site of Cempoala, Veracruz."

On the whole, the Chanal phase exhibits evidence of Tula-Mazapan influences, but it is difficult to know how fundamental they are. Unfortunately, the Chanal material is imperfectly known. All of the spectacular specimens removed from the cemetery west of the Río de Colima went immediately into the hands of dealer-collectors (some of whom, it is said, presented themselves every afternoon, with full purse and accompanied by armed guards). We cannot even be certain that Plumbate—which is a trade ware, critical chronologically—came from the West Chanal cemetery. At Tula, it occurs "in all levels except the deepest" (Acosta 1956–57: 91) and apparently in great abundance (Cobean 1974: 37). I have seen no Plumbate from Colima, but the form of one black vessel attributed to El Chanal suggests it may be a copy. Moreover, some years ago, Dr. Harold McBride sent me several color slides of pottery he had photographed in private collections, including two effigy vessels reputedly from El Chanal, and both apparently Plumbate.

Otherwise, most of the ceramic traits selected by Dumond and Müller (1972: Fig. 4) as characteristic of the early Postclassic in central Mexico seem not to crop up in Chanal association, nor do most of the traits that Acosta (1956–57: 84, 86, Figs. 16–20) considers representative of Toltec ceramics at Tula. Moreover, nonceramic aspects of

the Chanal phase, so far as they are known, do not tie inextricably with Tula. For El Chanal there is no report of Atlantean stone figures, of "standard bearers," or of chacmools, and the "pyramids" with carved stone risers[15] are not specifically Toltec in style.

To summarize, the little information now available from Colima indicates a feeble penetration of Tula-Mazapan influences during Armería times, as evidenced by a few figurine fragments and vaguely Mazapanoid bowls (see Armería: External Relationships, above). Neither of these occurs in Chanal association, and the Colima situation suggests that the two elements in question represent an early strain within the Tula-Mazapan assemblage. To complicate matters, the Mazapanoid figurines from the west tend to be well made, with sharp detail, in contrast to the untidy specimens commonly attributed to the presumed hearth in Hidalgo. Moreover, it is possible that in the west the Mazapanoid figurines have a somewhat earlier concentration. This would accord with the fact that Classic dates there seem, unexpectedly, to be referable to several ceramic features that are considered Postclassic in central Mexico (see Comala: External Relationships, above). Some of our current notions concerning the Postclassic evidently are due for revision.

Apparently a more potent Tula-Mazapan infusion took place in post-Armería times and involved quite different culture traits. The two radiocarbon dates from the site of El Chanal (Table 1, nos. 2, 3) suggest a comparatively recent time level, for the earlier one is A.D. 1290. Some of the elements that appear in Chanal context seem in central Mexico to start in Tula-Mazapan and to continue into Aztec times—for example, the Chila-like braseros (Prof. Eduardo Matos, conversation), the great cylindrical Tlaloc censers (Cobean 1974: 35), and possibly the cut-out censers with two feet and handle. Any diffusion westward should be pre-Aztec; at least, as far as I know, no fragment of the typical Aztec black on orange pottery has come to light anywhere in the west, and Corona Núñez (1960c: 379) specifically reports a complete absence of Aztec material in the stretch of Michoacán coast that he inspected. It may be that the bloc formed by the Tarascans and their allies effectively cut off communication to the west.

In view of the late radiocarbon dates for Chanal, and the fact that some of the Tula-Mazapan traits at El Chanal seem to have survived into Aztec times, a late date must be set for the intrusion in Colima. The most obvious explanation might be a westward thrust immediately following the collapse of Tula, in A.D. 1156 (date from Prof. Wigberto Jiménez Moreno, September 1974). At that time, there must have been turmoil and dislocation of peoples in the central valleys, and in the course of the confusion some group(s) with a Tula-Mazapan affiliation might have wandered or been pushed westward.[16] Actually, two such groups—or one that split in two parts—may have settled in Colima, for it will be remembered that the Chanal phase does not spread mantlelike over the whole area, but is limited to El Chanal proper and to an enclave in the lower Coahuayana valley (see Dating and Distribution, immediately above). Unfortunately, there seems to be no recorded tradition of such a westward movement (Kirchhoff 1961).[17]

The suggested late intrusion could explain why the Chanal phase includes several diagnostic elements of diverse origin. For example, variants of Autlán polychrome have a distribution[18] that suggests a local stylistic development. Along with such essentially western traits, intrusive ones appear. One may be Plumbate. Others may be censers and braziers. These presumably stem from central Mexico, and ties with the latter should also account for masonry "pyramids," the urban center at El Chanal, and the elaborate funerary offerings that suggest socio-economic-religious stratification.

These must have been times of cultural ferment and innovation in Colima, for the Chanal phase is rather more than a felicitous fusion of two major ingredients; a third one must be assumed in order to explain knowledge of metals. Despite excavation at Tula over the years, virtually no metal has been found (Acosta 1956–57: 94; personal communication from Prof. Eduardo Matos), except a few bits referable to trade or to post-Toltec occupation. Accordingly, the Tula-Mazapan phase is not a likely source of metal working, which generally is assumed to trace to South America. As usual, timing is uncertain. Copper may be earlier in coastal Nayarit than in Colima; Meighan believes it abundant at Amapa by A.D. 900 and suggests that it may even go back to the end of the "Classic" in west Mexico (Meighan 1974: 1259).

In other words, it may be said tentatively that the Chanal phase seems to rest on a local, west-Mexican base that has assimilated certain elements of Tula-Mazapan culture and, from another extraneous source, proficiency in working metals. It remains to be seen to what extent the above suggestions will hold, once more information is available.

The Tula-Mazapan imprint in west Mexico is still poorly known, except for Mazapanoid figurines and a few specimens of Mazapanoid pottery, plus a weak scattering of Plumbate. However, for Amapa (Nayarit), Bell (1971: 707, 709) speaks of "frying-pan censers and appliquéd plainware censers of a type which date from the Postclassic in the Valley of Mexico." Until the Amapa site report is published,[19] we shall have little more than general statements concerning central Mexican influences that "appear strongly at Amapa (Cerritos phase) and are very evident at Peñitas (Mitlan phase)." In addition, the "two late Postclassic cultures of Nayarit, Amapa and Ixtlan, are closely related, but Amapa continues to show stronger central Mexican influence than has so far been revealed at Ixtlan" (Bell 1971: 750, 752).

Periquillo

The last phase to be considered is the Periquillo, which centers along the lower Armería, between the modern settlement of Periquillo and the Pacific. A thin, somewhat isolated occupation is evident some 20 km to the northeast, on the great cerro that bounds Tamala (on modern maps: Aquiles Serdán) on the west. There, natural benches on the upper part of the east-facing slope have been modified and long lines of bounding stones added. As far as I am aware, this is the nearest approximation to masonry in Periquillo context.

There is little information concerning mortuary practices. My small tests produced no burials, and monero

information is slim. From the site near Tamala is described a pit that undercut a natural stone ledge to form a cavity in which the body was deposited. A stone slab set in front suggests the stone that covers the entrance to an Ortices-Comala tomb. One Tamala monero reports having found the bones of a small child within a pottery olla.

Ceramics

The preponderant pottery is red, but there is also black to gray. Red ware often takes the form of fairly large, incurved vessels, whose rim characteristically has a "crochet-hook" profile. Another red vessel, which might be called a "vat," is extremely large, excessively thick and heavy, and open mouthed; base form is uncertain but perhaps slightly pointed. There is a comal with red-slipped floor, and on record is one fragment of a trough dipper but apparently none of a slant-mouthed dipper or of heavy, arched handles.

The one painted ware, Periquillo red on cream, is an open bowl, with decoration on the inner walls; occasionally a winsome, modeled human(?) head projects above the rim. The bowl has tripod supports, almost always solid; some are rather stubby and subconical; others are longer, with a couple of protuberances that suggest residual eyes. Feet are usually weathered, so that painted ornament is indeterminate. The interior walls of the bowl are cream-to-gray-to white-slipped, with simple parallel lines —straight or wavy—in red overpaint. The result is basically orange on a very pale ground and, despite the linear motifs, not suggestive of Mazapan. Engraving or incising in combination with such painting is rare, as is any special preparation of the floor for grinding.

A cluster of unique pottery vessels—black, or red—is being called Periquillo embellished, to borrow a term from Meggers, Evans, and Estrada (1965). The common form is a cántaro with a globular or pear-shaped body and a short vertical neck, which in some cases flares slightly at the rim. Small canteens, with the body flattened fore and aft, are also known (cf. the similar form mentioned above, under *Chanal: Ceramics*), but they do not share the decoration of the Embellished wares. The latter vessels often have the neck adorned with small, hollow circles, apparently reed impressions. Sometimes there is a human face in relief on the neck, or a human or animal (always a lizard?) figure on the body. Deep parallel horizontal grooving or fluting, combined with appliqué pellets, is characteristic. A typical specimen appears in Kan, Meighan, and Nicholson (1970: Fig. 187).

These embellished wares were assigned to the Periquillo complex on the strength of surface association at a large site in the El Columpio–La Báscula area, just downstream from modern Periquillo, on a series of river terraces on the east side of the Río Armería. Over the years, several colleagues have doubted that Periquillo embellished belonged in a phase contemporaneous with the Postclassic (Kan, Meighan, and Nicholson, cited above, express such uncertainty in the figure legend). However, the two stratigraphic tests made in the shallow deposit at El Columpio confirm Periquillo association. Neither test has much depth (Test 1, surface to 60 cm; Test 2, to 120 cm), and both are Periquillo phase throughout. In Test 1, Periquillo embellished occurs in the three uppermost cuts and is absent in the fourth (bottom) cut, which, in any case, produced almost no diagnostic pottery. In Test·2, with seven cuts, Periquillo embellished shows up in the three uppermost cuts; in the fourth, there is one fragment; in the lower levels, none. Not only does the embellished material occur with other Periquillo wares, but it seems to be rather a late development within that assemblage and eventually may prove the basis for a separate ceramic phase.

A perplexing matter—and one by no means confined to west Mexico—is the "early" appearance of certain elements that seem to go into cold storage for centuries, only to crop up anew in later association (see Chapter 3: *Distribution: Nayarit and Sinaloa;* Chapter 4: *Northwest South America: Resemblances;* Tolstoy 1958: 67; Rattray 1966:188; Chadwick 1971a: 242; 1971b: 672; Dumond and Muller 1972: 1213–15; Menzel 1973). West Mexico provides numerous examples of this phenomenon—for instance, the modeling of a human or animal figure on the belly or neck of a vessel (cf. Ford 1969: 145); at least one such case is known from the Ortices phase (Kelly 1978: Fig. 15) and another, in east Colima, from a phase still unplaced but presumably Preclassic. In spite of relatively "early" occurrence in Ortices, in several parts of west Mexico such modeling appears in comparatively "late" company. One instance is from Culiacán, Sinaloa (Kelly 1945a: 104–5, Fig. 56); another is from Apatzingán, Michoacán, where raised figures on the belly of a cántaro occur in Llano red, a component of the Chila phase (Kelly 1947: Pl. 10 *b*). Still another example is provided by the embellished ware of the Periquillo phase of Colima.

Two well-defined figurine styles are Periquillo products, and fragments of both occur in surface collections and in the tests at El Columpio. One has an elongated slab body, to the top of which is affixed a transverse slab to represent the head; arms and feet are scarcely shown (Fish 1974: Figs. 1 *a–c, e,* 2). The whole thing is painted in the orange and off-white color scheme of the Periquillo bowls mentioned above. A specimen in the Diego Rivera collection has been published at least twice (Médioni and Pinto 1941: no. 77, attributed to the Sinaloa-Nayarit border, probably erroneously; and Museo Diego Rivera-Anahuacalli 1968: [p. 85], the latter specimen in color and evidently photographed following minor restoration).

The other figurine style is more varied and, it would seem, always red slipped. Excellent examples are illustrated by Eisleb (1971: nos. 162, 163), and chances are that a specimen attributed by Médioni and Pinto (1941: nos. 75, 76) to the Sinaloa-Nayarit border (like the one noted above) may be an atypical Periquillo red figurine.

Metal

Three items attest the presence of copper in the Periquillo phase: a small bell found on the surface of a monero dump at El Columpio, and another bell and a boss fragment from my tests at the same site.

Moreover, Periquillo attribution of a purchased lot of copper objects allegedly from the site near Tamala (mentioned earlier) appears reliable. I visited the scene with the moneros a few weeks after their "excavations" and observed bone fragments and bone beads(?) discarded on the surface and still bright green from contact with the copper. The lot includes eyed needles and several bits of wire, some in the form of circlets, as well as 65 bells, at least two with simulated wire work and with a Tlaloc face(?) on the resonator (cf. *Chanal: Metal*). A small chisel with a squared head may be a Periquillo-phase artifact, and at a Periquillo site near the coast, a farmer volunteered casually that once he had found there a copper "face" (mask?) eroded to the surface.

No gold or silver can be ascribed to Periquillo times, although excavation for a latrine some years ago on the outskirts of modern Tecomán produced 150 bells which, when cleaned, were "gray." They came from the same corral, if not the same burial, as a small plate allegedly of gold. Association is unknown, but I was shown at least one unequivocal Periquillo vessel said to have been found in the same corral.

Stone

Obsidian tends to be rather scarce in Periquillo association, and the same seems true of manos and metates. A private collection in Guadalajara has a large number of nicely worked metates with simple geometric designs pecked on the exterior walls, attributed to the neighborhood of modern Periquillo. Such decorated metates come chiefly from eastern Colima as tomb furniture and perhaps belong to a local extension of one of the phases of the eastern zone, contemporaneous possibly with Ortices-Comala of the Armería axis.

Dating

The one radiocarbon date for Periquillo is susceptible of two readings, A.D. 1450 or A.D. 1600 (Table 1, no. 1), making a considerable gap between Armería and Periquillo dates. A few Armería sherds are scattered through my tests at El Columpio but probably have no temporal significance because the deposits were so shallow. Furthermore, in the lower Armería drainage, Periquillo and Armería sites occur cheek by jowl, yet surface collections show no admixture.

Partial—if not full—contemporaneity between Periquillo and Chanal must be assumed. There are a number of generic resemblances in ceramics: a comal; a cántaro with horizontal midbelly seam; a canteen, with body so compressed that, viewed from above, it is elliptical; an open bowl, tripod, with color scheme of orange on near white, and rarely with a specially prepared molcajete floor which, when it occurs, is diagonally cross incised. Of presumably ceremonial ceramics, Periquillo has Chila-like braziers, and in one private collection is a typical Chanal "singing figure," said to have come from excavations in the modern town of Tecomán. Copper is securely attributed to the Periquillo phase, but there is some doubt concerning gold and silver.

There also are differences in inventory. As far as is known, Periquillo has no arched handle; no slant-mouthed dipper; no columnar Tlaloc "urns"; and no open-work censers. In fact, most of the Tula-Mazapan traits mentioned above for Chanal (under *External Relationships*) are absent, and it may be added that the latter phase lacks the handsome Periquillo embellished pottery, perhaps the oversized "vats," and certainly the typical Periquillo figurines.

In distribution, Chanal and Periquillo materials do not overlap physically but, as noted above (see *Chanal: Dating and Distribution*), there are sites in intermediate areas whose surface sherds include comales but whose pottery, on the whole, seems neither Chanal nor Periquillo. Several small sites on the coast, likewise, must belong in the same general time slot, but without being either Chanal or Periquillo. The temporal and regional differences that account for such variables await clarification.

It is not possible to identify the great El Chanal site with any sixteenth-century pueblo, and the same holds for the one really extensive Periquillo-phase site at El Columpio–La Báscula. The latter, in spite of its radiocarbon date and its close proximity to modern Tecomán, cannot be the sixteenth-century settlement of Tecomán, for the latter was said to be a quarter of a league from the sea (Suma de Visitas [Paso y Troncoso 1905: no. 681]; Sauer 1948: 49), whereas El Columpio must be a good 10 kilometers (presumably 2.5 leagues) inland.

External Relationships

On the score of Periquillo relationships with other areas, little can be said. Some resemblances to and differences from Chanal have just been noted; these seem meaningful but cannot be seen in much perspective until more is known of the content of both phases.

What promises to be important is the apparent lack in Periquillo of certain traits such as the slant-mouthed dipper and the arched handle; such absences are unexpected because these elements seem widespread in Colima and outlying areas at the time level of Periquillo. It may also be significant that most of the Tula-Mazapan traits found in the Chanal phase are not reported for Periquillo; however, these are known chiefly from the cemetery of West Chanal, and if a similarly well-stocked Periquillo burial ground were excavated, they might well appear.

In our present state of ignorance, Periquillo gives the impression of being a late phase, overlapping in time with Chanal, yet rather isolated culturally if not geographically.

COMMENTS

The foregoing data change considerably the previously published list of phases for the Armería axis. Capacha has been added at the early end, carrying the series far back in time; Ortices has been divided into the Ortices and Comala phases; and Chanal now appears as a separate entity. By no means do these changes take care of all current problems, and somewhere on the shaft-tomb time level are two potential phases still floating—ill defined and unplaced.

One such phase, tentatively called Manchón, is represented by pottery similarly named. It is known principally from 10 restorable or semirestorable vessels from a tomb near Los Ortices (Kelly 1978: Figs. 22, 23). Most of the specimens are red and close to utilitarian, but a few have special surface treatment that gives an apparent red on brown effect, in one case with slightly raised vertical ribbing. From this same tomb come a number of Ortices sherds attesting its initial use during that phase. Subsequently, the Manchón offerings were deposited, but later they were pushed to one side (and broken) to make way for one or more Comala-phase interments (Kelly 1978). The Manchón pottery appears not to belong in the ceramic inventory of the Armería axis. It may center in the upper Salado drainage and downstream, near Ixtlahuacán, at a time corresponding to that of the Comala phase, but the suggested chronological position needs substantiation.

A second potential phase early in the series is called Parranda, because the few specimens of known provenience are attributed to La Parranda, site B, near Comala. The lot is represented by several small cántaros, almost cylindrical, some with walls slightly concave, a body form reminiscent of certain Chupícuaro vessels (Porter 1956: Fig. 4 *a*); however, the specimens from Colima have a comparatively restricted opening, with flaring neck. Decoration is limited to broad, vertical stripes of red or of red flanked by narrow purple-black lines. It is likely that some of the unclassified black on red cántaros with maze motif belong with the Parranda material, and Doña María Ahumada de Gómez calls attention to the similarity of the black paint. The same paint, also applied in a maze design, occurs on one of two fine hollow effigies in a private collection. Almost certainly they affiliate with Parranda, and their mint condition bespeaks tomb provenience. In any event, the purple-black paint of the potential Parranda phase resembles that of Ortices polychrome and suggests a pre-Comala placement. "Jars" of "purple on red" and "purple and red on buff" are mentioned for the Morett site and make it likely that, with fuller information, Parranda may be linked with the early occupation at that coastal site, dated 300 B.C. to A.D. 100 (Meighan 1972: 8, 18, 48–49).

There are, of course, numerous problems not yet sufficiently formulated to be included here. At least, it is hoped that the preceding synopsis may provide some concrete perspective and serve as a basis for discussion and further investigation. It has been prepared as a background for the Capacha phase, to which the balance of the present report is devoted.

3. THE CAPACHA PHASE

CAPACHA EMERGES

In 1966, after a lapse of many years, I resumed archaeological work in Colima. During that season, the factotum of the hotel in the city of Colima offered for sale fragments of what evidently had been a large monochrome vessel, open mouthed, cinctured, and decorated with simple broad-line incision and punctation (Figs. 17a; 19d). Although surface collections already amounted to several tons, only two sherds similar to this vessel had been observed. They came from the surface of a cemetery near Los Ortices, where numerous free burials had been sacked by local moneros (Appendix IV-B-unnumbered site, Mesa del Salate, 6321, 6322).

At that time, I had no idea where the proffered vessel and the orphan sherds might fit in the Colima scheme, but each time the survey was extended to a new area, local residents were asked if they had seen such pottery. Because of its form, a waisted vessel is known locally as a *bule,* or water gourd; owing largely to heavy firing clouds, the surface is predominantly dark; accordingly, the question focused on *bules negros.* It turned out that in one area such bules are called *jarrones.*

I was told of two tombs near Las Borregas that had produced black cinctured vessels and a black mask, which latter suggested the Comala phase, and not until the end of the 1968 season was there an unmistakably positive reply, when the survey was extended to Buenavista, near the headwaters of the Río Salado. There, when the question concerning bules arose, the assembled residents burst into laughter. It turned out that, a few years before, several of those present had collaborated in looting a cemetery at nearby El Barrigón (Fig. 1, no. 7), where a number of shaft tombs had been discovered. Moneros converged on the spot, hoping to find more tombs with figurines that would bring a good price. No further tombs were discovered but the cemetery was "excavated" thoroughly. It seems to have produced burials of different phases, but most of them turned out to be of Capacha affiliation. The bules that resulted had no market value, hence the vessels were broken on the spot or taken home for the children to use at Christmas as piñatas. A visit to the spot confirmed the presence on the surface of a few small sherds evidently related to the incised and punctate bule that had been purchased in 1966.

At the start of the 1969 season, I tested at El Barrigón (Appendix I, no. 7), but monero predecessors had dug so effectively that we found little except disturbed soil, odd, unarticulated bone fragments, and a relatively small number of sherds. Eventually, we located one skull fragment (Appendix V, no. 8632; Figs. 46, 47), but the rest of the skeleton had been within the range of monero operations. Upright, and held in place by a few small stones, a nearly entire bule (Fig. 19 b) was pressed against what once had been the face of the corpse.

Toward the end of the 1969 season, I visited La Capacha (Fig. 1, no. 6), a few kilometers northeast of the city of Colima, and a local resident produced several pottery specimens he had found a few days before. Included was a small Capacha-style bule (Figs. 16 e; 17 f) and a number of items reputedly found with it. The informant took me to a nearby field, where he had a pit partially excavated, and I stood by while he removed fragments of a monochrome olla (Fig. 11 b) and a bowl (not illustrated), neither decorated. In addition, a bird effigy (Figs. 13 f; 26 f) and a bule (Fig. 19 c) were extracted. Later, from the same pit came a jar (Fig. 26 b) with shallow finger grooving on the upper part of the body. All the specimens are what eventually came to be called Capacha monochrome and indicate clearly that products of that phase were not limited to bules.

These pieces from Capacha, together with the bule from El Barrigón, first suggested ties with the "Tlatilco style" of central Mexico (see Chapter 4: *"Tlatilco Style"*). About the same time, I realized that two vessels and a few sherds found near Apulco, Jalisco, (Fig. 1, no. 1), and reported years ago as unplaced in the Tuxcacuesco series, doubtless related to Capacha (Fig. 30 e; Kelly 1949: 83–84, 210 [no. 100]; Fig. 59 b, c, Pl. 14d). In short, the Capacha phase began to take form.

Late in the spring of 1970, I tested the cemetery at La Capacha (Appendix I, no. 6) and found that there, as at El Barrigón, the moneros had exhausted the site. Disappointed, I went to Comala, to investigate the report of a site which reputedly had produced material of the presumed Parranda phase (see Chapter 2: *Comments*), not tied to the local sequence. As I waited to engage one of the workmen, I noticed the distinctive neck of a Capacha bule on the floor in a corner of his house. It proved to have come from nearby La Parranda and, within a few hours, we inspected the site, obtained the owner's permission to dig, and arranged for men to start work the next day. At La Parranda, site A, (Appendix I, no. 3), we found several burials, with a total of 11 bules. This was gratifying, particularly because bone fragments from some of the interments gave hope—futile, as it turned out—of a radiocarbon date. Grave furniture from La Parranda A consisted almost exclusively of bules, and it seemed advisable to learn what other artifacts could be attributed to Capacha.

Before we finished at La Parranda A, a monero mentioned having found similar pottery recently at nearby La Cañada. There, among the sherds he and his companions had discarded on the surface, were fragments of two restorable painted and incised vessels (Fig. 30 *b, bb, c, cc*). Subsequently, we tested several days at La Cañada (Appendix I, no. 2) and were fortunate in finding a number of burials that shed additional light on the Capacha phase.

The last bule cemetery worked in the 1970 season was at Quintero (Fig. 1, no. *10*; Appendix I, no. 10) near Caután, Ixtlahuacán, on the east bank of the Río Salado. Investigation in this area started with a survey designed primarily to try to establish cross ties between the Armería axis and the Naranjo-Coahuayana drainage of east Colima, for sherds characteristic of both occur on the surface. In the course of initial chatting with a monero-guide, I remarked that black wares seemed scarce. He agreed, but added that there *was* a local black pottery, of poor quality, in the form of bules. It turned out that the cemetery that produced this pottery was about 30 m from the informant's dwelling. We dug briefly at Quintero, finding several burials and additional Capacha ceramic forms, including a unique two-tubed stirrup vessel (Figs. 13 *d*; 25 *a, aa*).

Meanwhile, prior to the excavations at La Cañada and Quintero, I had gone to Mexico City to attend the 1970 meetings of the Society for American Archaeology. Several students of the Preclassic (David Grove, Louise Paradis, Paul Tolstoy, and Muriel Porter Weaver) were invited to look at the Capacha material then available. At that time, it consisted chiefly of bules, but all the group felt it belonged somewhere on the time level of the Mesoamerican Preclassic. In other words, for the first time there seemed to be strong suggestion of relatively early occupation in Colima. The balance of the present report is devoted to the description of the evidence and to a consideration of Capacha relationships to other areas.

Of necessity, a good deal of material included in this paper consists of tedious detail which, nonetheless, should be put on record. For reference, such data, along with the illustrations, are relegated to several appendixes; in the main text, salient features of the phase are selected and summarized.

The appendixes are as follows: I, Cemeteries and tests; II, Ceramics; III, Stone artifacts; IV, Capacha materials: provenience and associations. The latter appendix is intended to facilitate reference to all Capacha material in the present collection, be the association secure, alleged, probable, or possible. Appendix V, which concerns skeletal remains, has been prepared by Prof. José Antonio Pompa. Sparse and fragmentary though the Capacha skeletal material may be, it has suggestive and potentially important resemblances to "early" remains from other areas.

SITES

There is no surface indication of Capacha remains except at places where graves have been sacked and discards left on the surface. Monero activity has been destructive **and** would have been more so had there been a ready market for the comparatively unglamorous Capacha products. In one sense, indiscriminate monero digging has been helpful in that it has brought sherds to the surface; except for one site (Terreno de Jesús Gutiérrez), the Capacha cemeteries were brought to my attention by monero informants.

No occupational sites have been found, and the little information available comes exclusively from small cemeteries, relentlessly looted. Presumably these were close to habitations, yet remains of the latter have escaped detection. As a consequence, nothing can be said concerning the situation of Capacha settlements or the occurrence of artificial mounds or other features. In the absence of refuse, there is no information concerning food resources, but manos and metates attest some sort of grinding operations.

So far, there is no association of Capacha material with shell mounds, but future field work might modify this impression. Parts of coastal Colima have not been surveyed adequately. However, no shell, worked or otherwise, comes from a Capacha site—except several fragments of a single shell, apparently fresh-water, in company with mixed sherd material from general digging at the Quintero cemetery. As matters now stand, Capacha sites are not found on the low, coastal strip but are somewhat inland, at low to middle elevation—for example, Ixtlahuacán, Colima (186 m above sea level), and Tuxcacuesco, Jalisco (800 m).

Presently known Capacha-phase cemeteries are entered by number on Figure 1 (and described in detail in Appendix I). Information concerning them may be summarized thus:

Known through purchased specimens and surface collection: no. *1*.

Known through surface material alone: no. *4*. Two fragments on the surface of a rifled cemetery near Los Ortices (Appendix IV-B-unnumbered site, Mesa del Salate) have not been considered sufficient evidence to class this as a Capacha site.

Tested, and found effectively exhausted: nos. *6, 7,* and *8*; from nos. *6* and *8* some material was purchased.

Tested, with useful results: nos. *2, 3, 10*.

Tested, with results of very limited utility, but with definite evidence of Capacha affiliations: no. *5*.

Known exclusively from informants' reports: no *9*.

There is no obvious Capacha preference for a burial site, although proximity to settlement presumably was a consideration. Most cemeteries are near present arroyos or streams, but two (nos. *3* and *4*) are quite removed. One (no. *2*) is adjacent to a marked depression that once must have been the course of a modest stream; another (no. *5*) is on the bank of a perennial river, several meters above the level of the present stream bed. Some are on relatively high land, but one (no. *1*) is at the mouth of an arroyo and another (no. *10*) is on a sandy flat adjacent to the present flood plain of an important river. Three (nos. *2, 7, 9*) are in close proximity to shaft tombs, and indeed, some reports indicate that Capacha material has come from such tombs (no. *9*) (this chapter, *Interments*).

a. La Cañada (Fig. 1, no. *2*) (pp. 39–43).

b. La Parranda, site A (Fig. 1, no. *3*) (pp. 43–44). The Capacha cemetery is in the middle distance, where the figures of several men are just discernible.

c. El Barrigón (Fig. 1, no. *7*) (pp. 47–49). The Capacha-phase cemetery is on the flat, to the left of the low rise in the middle background.

[20]

d. La Capacha (Fig. 1, no. *6*) (pp. 46–47). The thorny brush is being cleared preparatory to testing.

e. Terreno de Fidel Valladares (Fig. 1, no. *8*) (pp. 49–50). The handsome stirrup vessel shown in Figure 25 *c* was exposed in the wall of the ditch in the foreground.

f. Quintero (Fig. 1, no. *10*) (pp. 50–53). Test in the bank of the old road, now an incipient arroyo. It is here that Burials 3, 4, and 5 (Fig. 8) were found. The main part of the Capacha cemetery lies behind the vegetation on the right. The present road is barely visible as a light area, left background.

Fig. 2. Landscape and excavations.

DISTRIBUTION

The distribution of presently known Capacha cemeteries is shown in Figure 1, but a number of pottery specimens, clearly of Capacha affiliation, indicate a considerably wider spread.

Jalisco

Presence of the Capacha phase at Apulco (Tuxcacuesco, Jalisco) is firmly established (Fig. 1, no. *1*) and is confirmed by a small but interesting assortment of Capacha vessels reported from still another cemetery near Apulco (Meighan 1974: Fig. 3; Meighan and Greengo 1974).[20]

One vessel (Fig. 24 *c*) thought to be from Autlán (Jalisco) was photographed 25 years ago in a private collection in Guadalajara. It seems clearly to be a variant of a Capacha compound vessel with trifid tubes. Autlán, slightly north of west from Tuxcacuesco (Fig. 1), is likewise in the Armería drainage. It has archaeological ties with Tuxcacuesco and, by extension, with Colima. The specimen in question indicates that Autlán-Tuxcacuesco-Colima cultural relationships go much farther back in time than had been suspected heretofore.

From Zapotiltic on the Colima-Jalisco border (Fig. 1), Schöndube (1973–74 II: lám. 66, no. 4) illustrates a figurine which he feels may have Capacha resemblances (personal communication). It is quite likely that this marginal area shared Capacha culture.

Nayarit and Sinaloa

As yet, there is no report of Capacha material from Nayarit, although effigy stirrup pots are suggestive (Chapter 4: *Nayarit*).

From Sinaloa, farther northwest along the Pacific coast and roughly opposite the tip of the peninsula of Baja California, there is an unexpected clue to Capacha presence in the form of several small trifid pots, whose tubes have the unique "elbow" of the Capacha stirrup-pot variant (see this chapter, *Ceramics: Form*).

Recognition of these vessels came almost by accident. In 1974, in the storeroom of the Museo Nacional de Antropología e Historia, Profa. Carolyn Czitrom noticed a small pot, almost a miniature. A trifid, with Capacha-like tubes, it appears to have had a thin red slip, and on its upper and lower bodies is a band of black with simple red and orange designs in overpaint (Museo no. 2.1-166, 58304; Fig. 30 *i*). Thoughtfully, she called the specimen to my attention, saying that Profa. Rosa María Reyna had remarked that this was one of several similar vessels. Later, Mrs. Reyna was kind enough to accompany me to the storeroom, where two other pots with similarly angled tubes were located. One of these (no. 2.1-78, 41055) is unusual in that the upper body is in the form of an incurved bowl, not a jar. It has vague red on cream decoration, but its "elbowed" tubes are emphatically Capacha. The third small trifid in the Museo (no. 2.1-74, 41050) has the upper body in the form of a jar—normal for Capacha—and has a red matte slip. Of the three Museo specimens, this is the only one with data of provenience; it formed part of a collection delivered to the Museo by Dr. Gordon Ekholm.

I wrote Dr. Ekholm immediately. In reply he stated that the specimen in question was purchased in Los Mochis, Sinaloa, in 1940, and was said to have been found in the vicinity of El Dorado, in the lower San Lorenzo Valley, Sinaloa. He sent a copy of the catalog that covers the entire purchased lot, together with photographs of a number of the specimens (personal communication, 23 October 1974). This same purchased collection evidently included a second near-Capacha pot, but fragmentary. Although the upper part is missing, the "elbowed" tubes are clearly visible in the photograph. This fragmentary vessel was released for export and is now in the American Museum of Natural History (no. 4793).

Because of their resemblance in form to the Sinaloan specimen delivered by Dr. Ekholm, the Museo's two small trifids without provenience have been cataloged as "probably from Sinaloa." On one (Fig. 30 *i*), someone has written "Michoacán," but I am assured that this attribution is not reliable.

In summary, the Museo Nacional has three small trifids with connecting tubes unmistakably Capacha in form, and there is an additional specimen in the American Museum of Natural History. Of the four, two seem certainly to come from the Sinaloan coast—an occurrence not easy to explain. Clearly, the form is Capacha-inspired, although it seems likely that dating may be much more recent. Unfortunately, the companion pieces in the purchased lot attributed to El Dorado are not helpful. Except for the two trifid vessels, there is little reminiscent of Capacha, although some handsome flat stone axes, three-quarter grooved, and with scrolls pecked on the face, might accord with Capacha skill in working stone (see Appendix III: *Ground Stone*). Temporally, the purchased lot seems mixed and includes polychrome, red on cream vessels, pipes, spindle whorls, cylinder seals, and quite strange figurine fragments, some of which have elaborate, almost turreted headdresses and excrescences each side of the head.[21]

The only explanation that comes to mind is that pre-Conquest excavations in the El Dorado area unearthed one or more Capacha trifid-spout vessels; the novel form was considered attractive and was copied. Whether or not the suggestion is correct, the potter(s) who made these stirrup-mouth variants surely had a Capacha model at hand, and, irrespective of dating, the El Dorado vessels provide unequivocal evidence of Capacha culture on the central coast of Sinaloa. Field investigation is warranted.

Otherwise, little reminiscent of Capacha is known from Sinaloa. Sauer and Brand's Tacuichamona material from the piedmont (1932: Fig. 14) is characterized by incision and punctation; inasmuch as it includes pipes, it is not likely to be "early." Heavy, undecorated sherds, some with rasped surface, were collected at sites along the coast at Culiacán (Kelly 1945a: 161–63, Fig. 77). The few published profiles do not look Capacha-like; deep, incurved vessels with thickening on the underside of the rim suggest an "early" placement, but non-Capacha.

Baja California

The presumed occurrence in the lower San Lorenzo valley of Sinaloa carries evidence of Capacha well up the Pacific coast toward Baja California, from an unspecified part of which comes the orphan stirrup vessel published by Brainerd (1949). Description and photograph indicate no specific Capacha features, but in our present state of ignorance, anything of stirrup-pot form merits attention. This isolated example from Baja California ends the long Capacha trail from Colima northwest along the Pacific.

Michoacán and Guerrero

Northeast, east, and southeast of Colima, an enormous expanse of country in the states of Michoacán and Guerrero is little known archaeologically. Chadwick (1971b: 659) lists excavations and surveys published for Michoacán and adjacent areas. In the highlands of that state, near Jacona, are the several tombs of the Opeño phase; because of demonstrable ties with Capacha, they are treated separately in Chapter 4 (*Opeño Phase*). The *tierra caliente* of Michoacán is known principally through rapid surveys and the limited excavations at Apatzingán (Kelly 1947); in passing, it may be noted that the latter's Chumbícuaro phase has a faint suggestion of Capacha traits.[22]

Part of the Michoacán-Guerrero boundary is formed by the Río Balsas. From this area and from western and northern Guerrero, there is a good deal of scattered information but no site report. A dozen or more surveys have been published briefly and with a minimum of illustrations. Indication of Capacha ties is slight, although from Santa Bárbara, south of Jaleaca, Guerrero, there is mention of "astonishingly thick coarse pottery . . . perhaps of great jars . . . from deep down in the side of an arroyo. This pottery was light yellow in color and grooved on the outside" (Weitlaner and Barlow 1944: 366). Seldom is Capacha "light yellow"; otherwise the description might apply to the noncinctured parts of its bules.

Archaeological salvage in the areas affected by dams at El Infiernillo and La Villita on the lower Balsas has produced some indication of "Formative" occupation (Litvak 1968; Chadwick 1971b: 660, 662, 665), but without recognizable hint of Capacha resemblances. The same is true of the provocative Pox pottery of Acapulco, Guerrero, which may date much earlier than Capacha (Brush 1965); more recent ceramics from that same area have a suggestive combination of grooving and punctation (Brush 1969), but neither form nor precise decorative treatment seems Capacha-like.

Guerrero is a large state. In its northeast corner, near the boundaries with the states of México and Morelos, a cemetery with "Tlatilco-style" pottery was excavated a few years ago. Ill-defined though this ceramic "style" is, it has some significant resemblances to Capacha (see Chapter 4: *"Tlatilco Style"*). The tri-state area in question still is a long way from the nearest-known Capacha remains and evidently relates to the strong "Tlatilco style" ingredient in Morelos. Accordingly, the prospect of establishing physical continuity between Capacha and the Tlatilco style still seems remote. However, Capacha figurines have some resemblance to certain specimens from Morelos; this fact, plus Capacha-Opeño resemblances, suggest that eventually some link across Michoacán-Guerrero may close the present gap.[23]

INTERMENTS

Information concerning Capacha skeletal material comes from fragmentary remains (Appendix V), poorly preserved and partially destroyed through recurrent use of the same cemetery. We found no long bone complete, and all bone was so leached it could not be used for radiocarbon dating.

At present, shaft tombs cannot be attributed to the Capacha phase, but several independent reports indicate that pottery—from description, Capacha—was removed from such tombs.

Three tombs in the vicinity of Las Borregas (Fig. 1, no. 9) seem to have produced Capacha material. Two were described by two brothers who had worked jointly. The third was reported by the excavator and confirmed, much later, by a chance witness; from this tomb, I saw two bules that had survived.

Furthermore, a monero in Comala remembered that, years before, he and his brother had dug two tombs at La Angostura, within the political limits of Suchitlán, but actually closer to Nogueras, a few kilometers north of the city of Colima. The chambers already had been rifled, he said, and the two men simply collected specimens left by previous looters. One tomb contained a black bule; the other, a black bottle, whose form was "like that of any bottle." I visited the scene, which is a barren, south-facing slope. Surface sherds were negligible in quantity and completely weathered; after the lapse of more than a decade, there was only faint indication of the sacked tombs. However, the precise spot was identified by the owner of the land, who had been present and who remembered the excavation in question.

The above boils down to reports of pottery, presumably Capacha, from four different tombs. It is likely that shaft tombs were a Capacha trait, especially since the phase has a few specific ceramic ties with Michoacán's roughly contemporaneous Opeño material, the latter known exclusively from tombs.

Monero information sheds further light on Capacha interments. Of several burials which had been excavated at the La Capacha cemetery, one was unusual in having three adjoining "compartments," roughly quadrilateral, cut slightly into subsoil. Two of the "compartments" had skeletal material—one, it was thought, with the skull to the west; the other, with the skull to the southwest. Reputed artifact association is given in Appendix IV (C-6, 8679–8687).

The same monero-informant had another excavation open when I first visited La Capacha, and I was present when the exposed furniture was removed (Appendix IV-A-6, 8676 a, b, 8678, 8706–8708). This burial pit was ill defined, roughly oval, its north-south axis about 130 cm, its east-west axis, 110 cm. On the south were two salients which it was thought might turn into additional "compartments," but this proved not to be the case. The tepetate floor of the pit containing the interment was extremely uneven and bumpy; at the north, it came within 30 cm of the surface; at the south, 90 cm. Several tiny bits of bone were found, but not enough to give any idea of orientation of the body.

A second monero who dug at the La Capacha cemetery reported having found a "trench" cut into tepetate, about

50 cm deep and 3 m long. Bones were accompanied by a Capacha bule and two other vessels (Appendix IV-C-6, 8712–8714). The same "trench" contained other Capacha material but no additional bone (Appendix IV-E-6, 8715–8720).

In addition to the monero information cited above, available data come largely from the small cemeteries at La Cañada, La Parranda (site A), and Quintero (Fig. 1, nos. *2, 3, 10,* respectively). The photographs in Figure 2 give an idea of terrain. A summary is provided by Table 2, and Appendix I presents basic information concerning cemeteries and excavations, together with sketches and photos of some of the interments (Figs. 4–8).

Capacha interments cluster in small burial grounds which evidently had recurrent use. Interments are extended and, as reported by moneros, some are within a shallow pit cut into subsoil. At Quintero, the skull consistently was to the north; elsewhere, orientation varied. Three prone burials—one apparently with the hands tied behind the back—probably are of the Capacha phase, but identification is conjectural, for none had offerings. However, such position is infrequent in Colima during other phases, and it may be no coincidence that "face-down" burials are common at the site of Tlatilco (personal communication from Dr. Muriel Porter Weaver, 6 August 1970).

A single case of secondary interment is known, with the fragmentary remains of two adults deposited within a bule at La Cañada, Burial 5 (Fig. 5, no. *5;* Table 2; Appendix V, no. 10293 a, b). A group burial consisted of a confused heap of bones; a few were articulated, but most were not. They were in a shallow pit in sterile gravel at a Capacha cemetery (Terreno de Fidel Valladares, Group Burial 2 [Appendix I, no. 8]); a few sherds of Capacha monochrome were within the pit, but no furniture. Another interment has been called a group burial (Table 2, La Cañada, Group Burial 2) because parts of two skeletons were involved, but this seems a case in which one interment disturbed a previous one.

The bule is the most common grave furniture. Group Burial 2 (Fig. 4) at La Cañada had two bules at the knees; a bule fragment near the pelvis may belong to a previous interment, disturbed when Burial 2 was placed in the prepared pit. Burial 4 (Fig. 5, no. *7*) of the same cemetery had an incurved bowl with painted ornament (Fig. 28 *f*) where the left femur should have been; against and partially overlying this vessel was a large bule (Figs. 5, no. *8; 17 e*). At La Parranda A, Burial 3 apparently had a bule at the head and a large Capacha olla at the feet (Fig. 6). Burial 1 (not illustrated) at the Parcela de Luis Salazar (Fig. 1, no. *5*) had an olla containing a miniature vessel at the left side of the thorax. Burial 10 at Quintero (not illustrated) had the skull partly within a basin of Capacha monochrome. In other cases, relationship of vessel to skeleton is uncertain owing to the fragmentary condition of the bone.

Sometimes specimens that were once grave furniture occur loose in the fill, as the consequence of churning during successive reuse of the plot. Many burials evidently have been disturbed. For example, it is likely that when the bule burial at La Cañada (Burial 5, Fig. 5, no. *5*) was

introduced into the prepared pit in subsoil, it dislocated Burial 3 (Fig. 5, no. *2*). In fact, the distribution of bone fragments suggests that more than one earlier burial may have been involved. It is odd that almost no bone came from the prepared pit at La Parranda A (Burial 4; Fig. 7, nos. *1–8*), which produced the largest number of restorable bules—seven, in fact.

There is no evidence that cemeteries were walled or otherwise were specially prepared. However, a protective layer of mud, mixed with small bits of limestone, apparently was spread over the interments at La Cañada. Sometimes large stones are associated with Capacha burials (Figs. 4–7, for example). Seldom are these slablike, and in no case are they placed at an angle over the body or laid horizontally upon it. Several horizontally arranged stones over the skull fragments at La Parranda A, test 5, probably are not of the Capacha phase, nor is the stone "pavement" uncovered at La Cañada (see Fig. 5).

The three principal Capacha cemeteries mentioned here —La Cañada, La Parranda A, and Quintero—are predominantly, but not exclusively, Capacha burial grounds. At La Cañada, a shaft tomb of the Comala phase was found on the western limits of the site, and several purchased artifacts attributed to the same cemetery are not Capacha (Appendix I, no. 2: *Comments*). La Parranda A likewise had several clearly non-Capacha burials (see Fig. 7). Quintero seems to have had at least one post-Capacha interment, represented by two large rose-slipped sherds flush with the surface; the rest of the furniture and the presumably accompanying skeleton evidently had been carried away by the adjacent arroyo (Appendix I, no 10: *Burials 1–6*).

CERAMICS: CAPACHA MONOCHROME

Pottery is described in Appendix II; here, only salient points are mentioned. The chief ceramic product is of thick, heavy, somewhat grainy paste and is called Capacha monochrome, despite the fact that a few variants are slipped or otherwise painted.

Form

The most characteristic shape is the bule (Fig. 9), a deep vessel with one cincture, seldom more, and with open, flaring mouth. Most specimens are relatively large and heavy. Form sometimes is reminiscent of the so-called Tlatilco style, but vessels of the latter type tend to be smaller, of better quality, and consistently with more restricted orifice. As far as I know, the bule is a local specialty, without precise counterpart elsewhere.

Capacha monochrome also includes cántaros and ollas. Some are comparatively large vessels; others remind one of the "small, wide-mouth pot" mentioned by Ford (1969: 86, Chart 12). Of bowls, there are various forms, most incurved, and some markedly so, but none with thickening on the underside of the rim. Occasionally the latter is pinched or indented (Fig. 12 *i*), and sometimes perforated, apparently for suspension (Figs. 12 *f–h, k; 22 f; 28 f*). One semirestorable hemispherical bowl (Fig. 22 *d*) has a squared rim on whose upper surface four small round punches have been made.

TABLE 2

Capacha Interments

| | Site Number (Figure 1) | | | | | | | | | | | | | | | | | | | | | | | | |
| | **2** La Cañada | | | | | **3** La Parranda (site A) | | | | **5** Parcela de Luis Salazar | | **7** El Barrigón | **8** Terreno de Fidel Valladares | | | **10** Quintero | | | | | | | | | |
	Burial 1[1]	Group Burial 2[3] (Fig. 4)	Burial 3 (Fig. 5)	Burial 4 (Fig. 5)	Burial 5 (Fig. 5)	Burial 2 (Fig. 3a)	Burial 3 (Fig. 6)	Burial 4 (Fig. 7)	Burial 5 (Fig. 7)	Burial 1	Burial 2[1]	Unnumbered	Burial 1[1]	Group Burial 2[1]	Burial 3[1]	Burial 1[1]	Burial 2	Burial 3[1] (Fig. 8)	Burial 4 (Fig. 8)	Burial 5 (Fig. 8)	Burial 6[1]	Burial 7	Burial 8[2] (Fig. 3b)	Burial 9[2]	Burial 10
Extended, head to north, face to west				X														X		X					
Extended, head to north, face to east or southeast																						X			X
Extended, head to north, face upward																X									
Extended, head to north, face sunk on chest																			X						
Extended, head to north, face down																	X								
Extended, head to north, face down, hands at sides																					X				
Extended, head to northeast		X																							
Extended, head to east, face down, hands together at back (tied?)	X																								
Extended, head to east, face to west										X	X														
Extended, head probably to northwest							X																		
Extended, or probably so, head to south												X	X		X										
Extended, head probably to south									X																
Double secondary burial within bule						X																			
Group burial[3]		X																							
Secondary group burial														X											
Too fragmentary for determination			X					X																	
Interment on or above subsoil	X								X	X	X	X	X			X	X	X	X	X	X	X	X	X	X
Interment in shallow pit but in subsoil		X	X	X	X	X								X	X										

1. Presumably Capacha phase, but allocation not certain, for want of grave furniture.
2. Presumed burials, but no accompanying bone.
3. Group burial in the sense that two skeletons are involved, but they probably represent successive interments, with the later disrupting the earlier (see Figure 4).

An assortment of stirrup-mouthed pots is of particular interest. Essentially, these Capacha products are vertically compound vessels, with upper and lower bodies connected by tubes (Figs. 13 *d, e*; 24 *b, c*; 25 *a–e*). Distinctively Capacha features are: the orifice in the form of a small vessel; a marked preference for three rather than two tubes; and a "jointed" or "elbowed" profile of the tubes, irrespective of number. Potentially, the stirrup vessel is of major significance in linking Capacha with other phases and other areas (see Chapter 4: *Honduras* and *Northwest South America*).

Furthermore, one stirrup pot is combined with a doughnut-shaped base, the latter a trait of potential significance, and especially so in combination with a stirrup-mouthed variant (Fig. 24 *c*).[20]

A few horizontally compound specimens, including small cántaros and plates (Fig. 13 *a–c*), also are Capacha in phase. There are suggestions of the narrow-necked bottle, but only one sherd (Fig. 10 *h*) might be from the neck of such a form. Flat-bottomed vessels, spouted trays, and molcajetes are absent, as are lugs, handles, and spouts of all kinds; tiny tripod and tetrapod supports occur only on miniature vessels (Fig. 14 *a–c, e, f*). The griddle, or comal, requires mention, although there is no assurance of its being a Capacha product. Theoretically, it could have been a service vessel which did not find its way into graves. Although paste, color, and texture are very unlike Capacha, a considerable number of comal fragments comes from Capacha cemeteries (see Appendix II: *Griddles*).

Decoration

Capacha pottery decoration focuses on broad-line incision and punctation, often combined in a highly distinctive manner, to form what may be called a sunburst (Figs. 15–17; 18 *a, b, d–k*; 19 *a, d*; 20 *c*; 21 *a–c*; 24 *a, b*; 25 *c, d*; 26 *a*; 29 *c–e*). Such ornament is found especially on bules, infrequently on other forms.

There is no rocker stamping and no cord marking. A few vessels have light fluting (finger grooving) (Figs. 26 *b*; 28 *b*; 29 *a*); raised ribs (Figs. 26 *c*; 27 *a, b*); or pronounced ridging (Figs. 26 *d*; 29 *h, i*) as decorative devices, and one specimen, almost surely Capacha, combines diagonal and horizontal ridging (Eisleb 1971: no. 192). An effigy vessel (Figs. 13 *f*; 26 *f*) and several miniatures (Fig. 14 *a–c*), the latter presumably Capacha, depend on three-dimensional modeling for ornament. In addition, a surprisingly sophisticated specimen, apparently a tortoise (Fig. 30 *b, bb*), is painted, incised, and modeled.

Painted Variants

A few painted vessels and fragments are attributable to the Capacha phase. Rather than set up a series of named types, each represented by one or two examples, it is preferable to treat these specimens simply as painted variants of Capacha monochrome. If new material warrants recognition of additional types, they can be added quite easily.

The painted vessels in Capacha association vary widely. There is no resist painting, but a rose-red slip (Fig. 29), red on brown (Fig. 30 *a–d*), and a combination of red and purple-black (Fig. 30 *e*) were probably more common than the present data indicate. The red on brown and the red and black combinations may have the design delineated by incision; in three cases, the zoning lines are filled with a white material that emphasizes the contrast in colors (Fig. 30 *b, bb, d, e*).

The red of the Capacha phase tends to be rose in tone —sometimes quite light, sometimes deep. Pending study by Ing. Adolphus Langenscheidt, it may be guessed that the pigment is hematite. A similar rose tone holds for all red paint used on Colima pottery prior to the Comala phase, and also characterizes some of the presumably "early" pottery from eastern Colima.

The purple-black paint is specular hematite (personal communication from Ing. Langenscheidt, 8 December 1971) and is represented in the Capacha collection by two sherds only. Similar purple-black pigment carries over into the Ortices and the presumed Parranda phases but it, too, drops out before Comala times. At least in the Colima area, hematite paint—specular and otherwise—seems to be a significant marker on the level of the Mesoamerican Preclassic. At Teotihuacan, specular red pigments do not appear until the Miccoatli phase (Teotihuacan II) (Millon, Drewitt, and Bennyhoff 1965: 34).

Summary and Comments

Capacha monochrome, together with its few painted variants, are the principal products known from the phase in question. All vessels originally accompanied burials, and the selection does not include the general run of domestic pottery, despite the fact that most specimens look utilitarian. There is a complete absence of food dishes and of fire-blackened vessels. Possible implications are considered in Appendix II (see *Unclassified and Trade Wares*).

Capacha monochrome has a few unusual aspects. Although waisted vessels occur in various parts of Nuclear America and at various time levels, the specific Capacha bule, with its wide mouth, is not known to me from elsewhere, nor is its ubiquitous sunburst decoration. To be sure, there are approximations in Tlatilco-style pottery (Fig. 30 *g, h*). A faint channel at the junction of neck and body of ollas and cántaros is a Capacha element (Figs. 10 *d, f, g*; 11 *h, i*; 27 *b*) that is shared with El Opeño (Fig. 11 *k*) and possibly also with Apatzingán.[22]

The stirrup pot—really a vertically compound vessel whose two bodies are connected by two or three tubes of highly distinctive profile—is a fundamental feature of Capacha ceramics (Figs. 13 *d, e*; 24 *b, c*; 25 *a–e*). The bottle may be absent, although a lone sherd (Fig. 10 *h*), rough on the interior, comes from a neck of a small diameter, and there is one report of a black bottle (see *Interments*, above). Two tapering, almost subconic necks (Fig. 10 *d*; possibly also 13 *a*; Eisleb 1971: No. 192) may or may not be considered bottles, according to one's definition. The incurved bowl is of particular interest because of its prevalence in early cultures elsewhere in the Americas, but, as noted above, no Capacha specimen has the underside of the rim thickened.

No supports—footed or annular—are attributable to Capacha except for those on miniature vessels, which include three tiny animal effigies. Nor is there any suggestion of molcajetes. Comal fragments occur at several Capacha cemeteries but may refer to a later time level. A few Capacha variants have a rose wash or simple geometric ornament painted in red on brown, occasionally with the motif zoned by incision. At least two of these painted and incised variants appear definitely sophisticated (Fig. 30 *b, bb, c, cc*).

In summary, Capacha is a sharply-defined ceramic phase, some of whose products are distinctive. Capacha monochrome is the earliest ware recognized so far in the Colima area and appears without known local antecedents. It disappears, leaving little in the way of perceptible heritage—except perhaps in the incised and punctate ornament of the tentatively defined Zapote and San Miguel phases of eastern Colima. Nevertheless, the external ties of Capacha pottery seem significant and are discussed in Chapter 4.

MISCELLANEOUS MANUFACTURES

Figurines are described in Appendix II. They vary considerably, but mention should be made of one fragmentary and three entire specimens (Figs. 32 *b*; 34) whose general effect is startlingly reminiscent of the Mesoamerican Preclassic. There is a strong suggestion of certain type D and type K figures (for example, Vaillant and Vaillant 1934: Fig. 6, Nos. *4, 11*; Fig. 8, Nos. *19, 21*). Several specimens (Figs. 33 *b*; 34 *aa, aaaa, bbbb, cc*) have a peculiar raised effect at the back of the head, as if the hair had been cut straight across; this detail is reminiscent of certain figures of central Mexico (Reyna 1971: 67, for D2 figures; for D2–C9, lám. 86, no. 10; lám. 87, nos. 1, 3; for K–D2, lám. 89, no. 4; for K *fina*, lám. 91, no. 10). In particular, the gopherlike mouth of Capacha specimens resembles that of well-known figures from Morelos (Grove 1970: Fig. 5 *l, m*).

Pottery oddments include whistles (ocarinas), beads, and perhaps one effigy pendant (fig. 35 *a–d*). A small lump of hematite (Appendix IV-B-8, 10227) is from one Capacha cemetery, and from another comes a small bit of clay with cane imprint (Appendix IV-B-2, 9811), presumably from some sort of wattle-and-daub structure.

Appendix III gives details concerning stone artifacts. Eleven obsidian specimens in Capacha association include small chips or flakes and one scraper (Fig. 36 *e*). A second scraper (Fig. 36 *d*) is probably Capacha, but the association of a retouched projectile point (Fig. 36 *f*) is suspect. If obsidian flakes were related to manioc preparation (Lowe, in Green and Lowe 1967: 58–59), they should be expectable at occupational sites rather than at cemeteries. Such small chips, incidentally, continue throughout the entire Colima archaeological sequence. Prismatic blades are not Capacha products, and probably the same holds for retouched artifacts. Obsidian dating is mentioned in the following section.

Manos and mano fragments (Fig. 40), some unused, have been found with burials. In addition, there are a number of metates and milling stones (Figs. 37–39), several well-worked receptacles (Figs. 42; 43 *a, b*), and, it would seem, axes with three-quarter groove (Fig. 44 *a*).

One small stone of fine black nonvesicular basalt (Fig. 44 *b*) resembles stones often considered polishers for pottery, although, at a guess, they could as easily be amulets. Prof. José Luis Franco (personal communication, 1973) says that such stones are common at Mesoamerican Preclassic sites in many parts of Mexico, especially at those with Olmec ties. Quite different kinds of hard stone are used, he says, but invariably the specimens are nicely worked. An occurrence in apparent Capacha context is consonant with the "early" placement of the phase.

CHRONOLOGY

Stratigraphy

There is no meaningful stratigraphy. No habitation sites are known, hence there is no rubbish, superimposed or otherwise. Most Capacha burials are shallow—some not 50 cm below the present surface—and interment was made in soil free of sherds or in areas where earlier Capacha burials were disturbed.

Several post-Capacha burials were found, but seldom in meaningful stratigraphic relation to the earlier interments. A "pavement" adjacent to Burial 4 (Fig. 5, no. *7*) at La Cañada probably dates from Colima-Armería times. To the same general time level belong two burials at La Parranda A, which actually overlie Capacha interments.

Two small fragments of unslipped Red unclassified ware came from La Parranda A, one immediately beneath the remains of the olla that accompanied Capacha-phase Burial 3 (Fig. 6, no. *6*). The position suggests the possibility of a still unknown pottery contemporaneous with Capacha or earlier (Appendix II: *Unclassified and Trade Wares*).

The sherd sample from the Quintero cemetery is noticeably more mixed than elsewhere, especially in the vicinity of Burials 8 and 10; it includes a surprising number of comal fragments. One large griddle sherd was found overlying a Capacha bule and almost resting upon it. These comales are not allocated to phase, but paste and finish are not Capacha-like (Appendix II: *Griddles*).

At present, there is no evidence of internal time differences within the Capacha phase.

Absolute Dating

Dating on the basis of obsidian hydration was attempted. A total of 77 obsidian specimens—mostly chips or flakes—comes from Capacha cemeteries, with provenience either certain or alleged (Appendix IV, under A-6, 7, 10; B-2, 3, 4, 5, 6, 7, 8, 10; E-6). Some clearly are non-Capacha; others are of mixed or uncertain association. At present, 11 obsidian specimens can be attributed securely to Capacha (see Appendix III).

When the phase first was recognized, 22 obsidian samples from Capacha cemeteries—association certain, reputed, or possible—were examined in the UCLA Obsidian Hydration Laboratory, through the courtesy of Dr. Clement Meighan and Mr. Leonard Foote. Only three specimens had been found in certain Capacha association, and it is to be regretted that subsequently excavated obsidian, firmly identified as to phase, was not submitted. The three specimens examined were not as useful as had been hoped: one had no visible hydration band; another showed a band of 8.4 microns (with a date of 234 B.C., at the conversion rate of 260 years the micron); still another was read as 8.5 microns (260 B.C.) (Meighan, Findlow, and De Atley 1974: 115, nos. 8634, 8676 A, B).

In contrast, a purchased lot of 11 chips (my Appendix IV-E-6, 8720 a–k; listed in Meighan, Findlow, and De Atley 1974: 112–13, nos. 1603–1613, as "Misc.") allegedly found with Capacha pottery, produced some of the oldest readings of the entire Colima series: two of 10.6 microns (806 B.C.), and one each of 10.4 (754 B.C.), 9.8 (598 B.C.), 9.6 (546 B.C.), and 9.5 (520 B.C.); the remaining specimens gave more recent dates.

Although there is a suggestive clustering of dates from several of these presumably Capacha flakes at the early end of the Colima obsidian sequence, it must be noted that the oldest of such dates (806 B.C.) falls half a millennium short of our uncorrected radiocarbon date of 1450 B.C. (see below). The latter, thought at first to be much too early, now seems well substantiated by evidence from outside the Colima area. Accordingly, I do not take the present obsidian dates literally but recognize merely that they place Capacha early in the local obsidian sequence.

During the 1973 season, a sample of charcoal in Capacha association was obtained but evidently was contaminated, for the resulting date fell within the historic period (Table 1, no. 15). Bone fragments were comparatively plentiful but so leached that they did not provide sufficient carbon for dating.

In view of these difficulties, a date based directly on the organic content of Capacha potsherds is the only one available at present. A considerable number of bule fragments were collected from the surface of site no. *4* (Fig. 1), where moneros had looted several interments. A generous sample of these sherds was submitted to Geochron, whose "pretreatment" was described as "roasting in oxygen to recover total organic content." The resulting date (GX 1784) was 3400 ± 200 B.P., or 1450 B.C. (corrected to 1870–1720 B.C.; Table 1, no. 14). This was far earlier than anticipated, but as additional material came from the field and as ties with early South American ceramics became evident, it began to look acceptable.

There is no indication whether this one Capacha date falls early or late within the life span of the phase. It may be pertinent, however, that Geochron (GX 1785) obtained a date of 3080 ± 190 years B.P., or 1130 B.C., from sherds of an east Colima ware, provisionally named Zapote rose. Another east Colima ceramic, tentatively named San Miguel monochrome, yielded a date of 1100 ± 220 B.C. when tested by thermoluminescence, courtesy of the University of Pennsylvania (P-T-440-A).

Fortunately, the Capacha date receives indirect substantiation from a radiocarbon date based on a charcoal sample from a tomb at El Opeño, Michoacán. The latter date (without adjustment) is placed at 1500 B.C. (Oliveros 1970: 135, 1974: 193, inexplicably "corrected" to 1280 ± 80 B.C. in Oliveros 1974: 197). Because of ceramic similarities between Capacha and Opeño (Chapter 4: *Opeño Phase*), the two independently derived dates provide mutual support for the early placement of both phases.

4. CAPACHA RELATIONSHIPS

MESOAMERICA

Capacha resemblances to Mesoamerican ceramics are slight and, it would appear, not very fundamental. To be sure, there are evident ties with the Opeño phase of El Opeño, in highland Michoacán, and with the ill-defined Tlatilco style, of the Valley of Mexico, Morelos, adjacent Guerrero, and Puebla. However, all three apparently cognate lots—Capacha, Opeño, and the Tlatilco style—seem fundamentally to have non-Mesoamerican affiliations (see this chapter: *Summary*). Otherwise, there is little basis for a comparison of Capacha with the strictly Mesoamerican ceramic assemblages that are more or less contemporaneous, and, in most cases, the patent lack of resemblance makes anything more than cursory mention beside the point.

Huasteca

Near Pánuco, far to the northeast of Colima, the Pavón phase includes two forms of cántaro and a flat-base bowl (MacNeish 1954: Fig. 19). The latter form is early and apparently basic in Mesoamerica. So also is the neckless jar, often called the *tecomate,* but it is lacking in Pavón and in the succeeding Huastecan phases. None of the shapes reported for Pavón occurs in Capacha. The latter's broad-line incision and punctation are absent in the Huasteca, while Capacha lacks the decorative overlapping punches of Pavón and the "grater-bowl" incision, which latter appears in the Ponce phase following Pavón (MacNeish 1954: Fig. 12, nos. 7–10, 1–6).

Southern Gulf Coast

To the southeast, in southern Veracruz, the Ojochi phase at San Lorenzo Tenochtitlan is a near Capacha contemporary. For it, Mesoamerica's two highly characteristic early forms are reported: tecomates and flat-bottomed bowls, with the former dominant (Coe 1970: 22). The presence of narrow-necked bottles is of special interest. Ojochi decoration centers on fluting, gadrooning, and simple red on brown stripes and bands; except for "stick punching" (Coe 1970: 22), there is no mention of incision or punctation. It is possible that the bottle occurs sparingly in Capacha context, but otherwise the Ojochi assemblage has little suggestive of Capacha.

The same is true for the Bajío phase, which follows Ojochi. Bottles continue, now abundantly, with body fluted or gadrooned (Coe 1970: 24). From the description, a neck "slightly constricted toward the mouth" suggests that of a Capacha specimen (Fig. 10 *d*; probably also Eisleb 1971: no. 192). Punctation, some zoned, sounds familiar, but the flat-base pan, rocker stamping, nail gouging, and differential firing are all remote from Capacha norms.

Considerably later, in the San Lorenzo A phase, pottery "which can only be called Olmec" appears and includes "carved" or excised ornament (Coe 1970: 26–27). Colima seems singularly free of Olmec contamination, although general digging at La Cañada (Fig. 1, no. *2*) produced one carved or excised sherd (Fig. 31 *k*) and the Terreno de Fidel Valladares (Fig. 1, no. *8*), another (not illustrated). The chances are that the Capacha phase is (1) somewhat earlier than the Olmec manifestation and (2) physically out of its range,[24] because post-Capacha phases likewise are uncomplicated by Olmec elements. The two excised sherds just mentioned presumably represent trade with some unknown area that shared Olmec ceramics. Although both are from Capacha cemeteries, association is not exclusively Capacha (Appendix IV-B-2, 9815, Fig. 31 *k*; IV-B-8, 10219 a).

Soconusco

Along the Pacific coast, close to the Mexican-Guatemalan border, another early ceramic series has been established through work at Altamira-Ocos-Salinas la Blanca. The earliest phase—known from one mound only, at Altamira—is called Barra (Lowe, in Green and Lowe 1967: 56).

Its Cotán grooved red pottery is characterized by a squat tecomate with flattened base, although the underside of the rim lacks the thickening often considered a tecomate feature. Vessels are slipped, some with pigment whose base is hematite. Decoration, sometimes elaborate, is based on fluting and grooving. Although Lowe (Green and Lowe 1967: 97–98, Fig. 72) points out suggestive resemblances to northwest South America, neither in form nor decoration is Cotán grooved red Capacha-like. As for the hematite slip, the chances are that Capacha's rose-red paint has a hematite base, but definitive study is awaited.

Another Barra-phase ceramic is Petacalapa black, with a flat-base bowl and "vases, plus jar form of unknown shape" (Lowe, in Green and Lowe 1967: 100, Fig. 74). Its zoned punctation is of special interest, and Capacha, with its profusion of incompletely zoned punctation, competes with Petacalapa black in having "the earliest-known zoned-punctate decoration in Mesoamerica." Unfortunately, Petacalapa black is sparsely represented at Altamira and disappears at the close of the Barra phase, so that the broad range of the ware is unknown. Available information suggests that Capacha resemblances are negligible.

The Altamira-Ocós phase, which follows Barra, has the familiar Mesoamerican combination of tecomate and flat-base pan. Coe (1970: 22) sees such close similarity between Ocós and Ojochi that he views the latter as a "kind of country cousin," hence the comments made above concerning Capacha-Ojochi apply likewise to Capacha-Ocós. The whole Ocós roster of decorative devices—zoned "dentate, shell-back, shell-edge, plain rocker-stamping, cord-marking, fabric impression, iridescent stripes" (Coe and Flannery 1967: Fig. 8)—is foreign to Capacha and, indeed, to the whole range of Colima ceramics, irrespective of time. One detail may be worth noting. A specular "hematite polished slip" (Coe and Flannery 1967: Fig. 8) rather suggests the purple-black paint of two sherds—variants of Capacha monochrome—which owes its sheen to particles of specular hematite (Appendix II: *Capacha Monochrome: Painted Variants*).

Chiapa de Corzo

The early part of a sequence established for Chiapa de Corzo (Dixon 1959) is contemporaneous with Ocós (Coe 1961: 123), and, like it, shows marked affinity to material from the southern Gulf Coast. In fact, Lowe (Green and Lowe 1967: 75) views the situation in terms of an "Olmec population block astride the Isthmus." As might be expected, the tecomate and the flat-base bowl are reported for Chiapa I. It is odd—and probably significant—that neither Ocós nor Chiapa I has the "narrow-necked" bottle attributed to Gulf-coast Ojochi (Coe 1970: 22). In any event, neither form nor decoration of early Chiapa is remotely suggestive of Capacha.

Nayarit

Also in connection with Ocós, mention may be made of a ceramic phase reported recently far to the northwest of Colima, on the coast of Nayarit. Mountjoy (1970: 45) feels this material relates both to Ocós and to Morett, which latter is mentioned a few paragraphs below. In view of such dual resemblances, the chance of relationship to Capacha is extremely slight.

Two human effigy stirrup vessels attributed to Nayarit (one, reported in Kelly 1972: 28) are without association,[25] as is a third thought to be from Colima (Corona Núñez 1960a: [lám. 9]). As a guess, all three are "early," and their rusticity is reminiscent of a vaguely similar specimen from Honduras, likewise without specific relationships (Porter 1953: Pl. 14). These occurrences cannot be meaningful until more is known of the archaeology of the areas in question, but the stirrup-spout vessel promises to be highly significant in tracing cultural ties (see this chapter, *Northwest South America: Resemblances*).

Sinaloa

From the area of El Dorado, on the lower San Lorenzo river, there is strong suggestion of outlying Capacha material, which has been discussed in Chapter 3 in connection with the distribution of the phase.

Coastal Guerrero

Although most of the archaeological material known from the state of Guerrero is said to fit comfortably into a general Mesoamerican scheme (Lister 1971: 619–20), some of the pottery from the coast near Acapulco warrants special comment. The earliest ceramic identified there has its "secondary surface" pitted, hence has been called Pox pottery. Brush (1965; 1969: 97) points out that this surface treatment is shared by the early pottery from the Tehuacán valley. A carbon date based on shell, without adjustment, is 2440 B.C., for the Pox pottery, roughly a millennium earlier than Capacha.

Neither Pox nor the more recent ceramic material from the Acapulco area shows close resemblance to Capacha. The early Guerrero form seems to be predominantly an incurved vessel (Brush 1969: 57), and there is no approximation to a bule. At first blush, a combination of incision and punctation is suggestive, but specific resemblances to Capacha are lacking and the decoration in question appears to postdate Capacha.

However, a figurine head (Fig. 31 *n*) quite similar to specimens from the Guerrero coast comes from La Parranda A; it suggests Colima-Guerrero trade relations, although not necessarily during Capacha times; unfortunately, the association in Colima is unclear.

* * *

Although the ceramic assemblages mentioned above are presumably more or less contemporaneous with Capacha, there is little in them that suggests genetic relationship with the latter. Next will be considered the ceramic phases physically within Mesoamerica or on its fringes, but with evident cultural ties elsewhere. First, however, it would be well to liquidate two negative cases, relatively close spatially but not temporally. One is Morett, a site on the Colima coast, adjacent to the Jalisco border. It may be eliminated on chronological grounds, for "the bulk" of its "Early period ceramics" dates "between the first century B.C. and the first century A.D." (Meighan 1972: 13). Chupícuaro, a late Preclassic site in Guanajuanto (Porter 1956; Natalie Wood Collection 1969) also is too recent.

Opeño Phase (Michoacán)

Opeño is the earlier of the two phases which Oliveros (1970, 1974) recognizes for the tombs at El Opeño, near Jacona, in highland Michoacán. The more recent phase, the Jacona, is not pertinent to the present theme.

Between the Opeño and Capacha phases there are only a few resemblances, but they are highly significant. Uncorrected radiocarbon dates indicate essential contemporaneity (Capacha: 1450 B.C.; Opeño: 1500 B.C.) and provide support for situating both at a time level far earlier than had been anticipated.

Ceramics

In form and decoration Capacha-Opeño resemblances are evident. Apart from suggestive similarities in general contour, an occasional Opeño vessel could be incorporated into the Capacha series without attracting notice—especially the one shown in Figure 11 *k,* which has the slight, highly distinctive Capacha channel at the base of the neck.

Specific Capacha resemblances are found also in the painted and incised group that Oliveros calls "Opeño Rojo y Guinda Esgrafiado" (Oliveros 1970: 83–87). This lot varies considerably. Zoning may be incised or engraved; *guinda* ranges from light rose to purple black, and the natural base ware may substitute for the red zones. One sherd, which combines rose with natural brown and has incised zoning enclosing a triangular area filled with round punctations (Oliveros 1974: Fig. 11, left, second from top), is very similar to a decorated variant of Capacha monochrome (Fig. 31 *h*). Although Capacha punctation is not characteristically circular, this kind is by no means unknown (Fig. 18 *a, c*).

Several other Opeño fragments (Oliveros 1974: Fig. 11, right, first and second from top) are obviously close to Capacha. Although classed as red and *guinda,* in this case the latter color verges on purple black, and if the bounding incision were filled with white, decoration would be exceedingly close to that of the once-unclassified cántaro from the Arroyo de San Antonio, near Tuxcacuesco (Fig. 30 *e*; Kelly 1949: Pl. 14 *d*). The latter specimen, now recognized as a Capacha product, was located in the storeroom of the Museo Nacional and an Opeño sherd placed against it; both Prof. Oliveros and I felt the two geographically discrete wares to be indistinguishable.

A painted and incised vessel excavated at El Opeño by Prof. Eduardo Noguera in 1938 is shown here as Figure 30 *f*; it is considered by Oliveros (1970: 85) a variant of the same group of wares mentioned in the two last paragraphs. The specimen in question might pass for a poor-quality example of a decorated variant of Capacha monochrome.[26]

A few Capacha and Opeño sherds are so similar that the possibility of trade was suspected, either from one area to the other, or from a common source. Accordingly, Prof. Oliveros and I submitted a small sample of sherds to Ing. Adolphus Langenscheidt and to Dr. Garman Harbottle, who studied the possibility of a common clay source according to their specialized approaches—respectively, microscopic inspection and activation analysis. Both investigations suggest—without, however, absolute proof—that Capacha and Opeño ceramics are products of their own respective localities (personal communication from Langenscheidt, 22 January 1971; Harbottle 1975).

Miscellaneous Manufactures

There is little obvious resemblance between Capacha and Opeño figurines.

In ground stone, a small tripod bowl (Oliveros 1970: lám. 15: 36) is not unlike a Capacha specimen (Fig. 42 *d*). Neither phase was bountifully supplied with obsidian, and prismatic blades were lacking. The presence of retouched projectile points is dubious, at least for Capacha (see Appendix III), although Oliveros (1970: Fig. 14:45) illustrates one apparently found in Opeño association.

Comments

The specific resemblances between the Capacha and Opeño phases are indicated above. To be sure, there are differences in ceramics and other products, and these far outnumber the similarities. It is impossible to guess to what extent the apparent likenesses and differences reflect no more than our scanty knowledge of both phases. In ceramics, Opeño has no bule, no stirrup pot, no incurved bowl with incised and punctate decoration, no sunburst motif. Capacha has no open bowl whose entire interior is covered with closely spaced punctation (Oliveros 1974: Fig. 11, bottom row). Nor does it have resist painting, which occurs on sherds from one part of an Opeño tomb that have been assigned to the Opeño phase (Oliveros 1974: Fig. 8:26, 27, 31–34).

Theoretically, the negative traits shared by Capacha and Opeño should be significant. As lacking in Opeño, Oliveros (1970: 108) lists the following, which he considers essentially Olmec: tecomates, kaolin ware, polished black ware, spouted trays, rasped or excised decoration, rocker stamping, Olmec figures of types A and B, and differential firing. He notes, furthermore, the absence of stirrup vessels, bottles, and seals.

The Capacha data agree, with some qualifications. Of the negative traits listed by Oliveros, Capacha does have a few incurved bowls, but without the thickening on the underside of the rim that often occurs on tecomates. Some Capacha sherds are polished, but its monochrome can scarcely be considered a polished black ware. Two sherds with excised ornament come from Capacha cemeteries but are considered trade products (Fig. 31 *k*; see also Appendix II: *Unclassified and Trade Wares*). Capacha certainly has the stirrup pot and variants, but the presence of the bottle is uncertain. As far as is known, seals and stamps do not occur at this time level in Colima.

To the preponderantly negative traits listed by Oliveros, a few more may be added: carinated bowls, flat-base pans, shoe-shaped pots, "grater" bowls, white slip, double-line break, and such typical Olmeca motifs as the were-jaguar, St. Andrew's cross, and what Weaver (1967: 38) calls the reversed E.

In summary, Capacha and Opeño share, to a considerable extent, the same negative elements. It would be gratifying if we could conclude that both phases were too early to have the traits in question, for that would substantiate the radiocarbon dates. However, with very few exceptions,[27] the negative traits listed are absent in west Mexico even in more recent times, so that spatial rather than temporal distribution seems involved. Perhaps the real explanation is that Capacha and Opeño—and, to some extent, the Tlatilco style as well—have a common heritage, apparently more South American than Mesoamerican, and a basic cultural dichotomy thus is reflected (see this chapter: *Summary*).

"Tlatilco Style" (Central Mexico)

Since 1947 the well-known site of Tlatilco, on the outskirts of Mexico City, has produced several hundred burials, accompanied by an extraordinarily rich, varied, and enigmatic assortment of grave goods. Certain of their ceramic elements have been called the "Tlatilco style" or the "Río Cuautla style" (Grove 1970: 67–68), the latter because many of the specimens from Tlatilco proper and related material from Morelos are indistinguishable.

There are suggestive resemblances between Capacha and some of the features attributed to the Tlatilco style, yet meaningful comparison is difficult because the "style" in question wants sharp definition and its temporal placement has fluctuated considerably.

Composition

Some years ago, the following ceramic traits were attributed to the Tlatilco style: red on brown pottery, stirrup bottles, composite bottles, belted bottles, and "composite botellónes" and "tubular-neck botellónes," (Grove 1970: 67–68), as well as the "ledged-neck" base (Grove 1970: Fig. 6 *c*; 1974b: 34–35). To this nucleus were added: gadrooning, masks, "certain effigies" (Tolstoy 1971: 26; 1973: 13), and whistling jars (Tolstoy and Paradis 1971: 17). All these traits are sufficiently distinctive to be recognizable readily, yet the "style" has been set up casually and without close analysis.

Recently, in fact, ideas concerning its composition have undergone drastic change, and Grove has written: "I would now call the Tlatilco complex stirrup spouted bottles, composite bottles, and bules basically" (personal communication, 29 September 1974). Specifically, he eliminates "tubular spouted bottles" and fails to mention most of the earlier-ascribed assortment of traits which presumably have been sloughed off.

Grave Lots

Interpretation of the heterogeneous material from the Tlatilco-site burials has been discussed over the years, and students have repeatedly noted the need for publication of the full lot of burial furniture. Grave associations, apparently incomplete, have been inspected by several investigators, with quite different conclusions.

Grove (1970: 70) declares that "grave lots containing the distinctive Olmec ceramics (excised wares, baby-face figures, spouted trays, etc.) did not contain Río Cuautla ceramics"—that is, he finds a clear dichotomy between Olmec and Tlatilco-style offerings.

Tolstoy's (1971: 26) incomplete seriation shows four "groups of burials," of which the fourth, "probably the latest, consists of graves containing stirrup-spouted vessels and other markers of the 'Tlatilco' style, as well as what appear to be occasional anticipations of Zacatenco ceramics."

Grennes-Ravitz (1974: 105) reports a still different situation. His comments concerning the Tlatilco-site graves are based on a compilation made by Sr. Luis Covarrubias of the furniture of 120 interments of "Temporada II." He notes that offerings "indisputably Olmec" are "in almost every case" accompanied by "Morelos material" (see below).

Moreover, viewing the Tlatilco-site burial furniture from the viewpoint of his Morelos data, Grennes-Ravitz (1974: 100) sees in it "at least three components": (1) one Olmec; (2) one related to later "Zacatenco" communities; and (3) one considered of "Morelos origin," and comprising the earlier part of his El Terror phase, which he calls La Manuela (Grennes-Ravitz 1974: 100, 102). This subphase includes figurines of D style, especially D2, plus K variants, red on brown bottles—globular, with vertical incision or light gadrooning—and other ceramics.

Grennes-Ravitz does not mention the Tlatilco style as such, but evidently it is what he identifies with La Manuela subphase.

In short, notions concerning the components of the Tlatilco style are in a state of flux and findings with respect to grave association vary. There also is difference of opinion regarding temporal position.

Chronology

Initially, Grove (1970: 71) considered the Río Cuautla or Tlatilco style post-Olmec and suggested it might be a convenient marker for the Middle Preclassic. Subsequently, he concluded that "the Tlatilco culture in the highlands is Early Formative and occurs generally within the Valley of Mexico's Ixtapaluca phase (c. 1350–900 B.C. in Morelos)" (Grove 1974a: 112). Of the Tlatilco style he says little specifically but has red on brown bottles appearing about 1250 B.C., with a great "complexity" of such vessels at roughly 1000 B.C., coinciding with the peak of "Olmec style" ceramics (Grove 1974a: 114).[28] He qualifies the statement by suggesting that "even though Olmec stylistic attributes and the Río Cuautla bottle complex appear in the same subphase, they need not have appeared at exactly the same time or have endured the same period of time" (Grove 1974b: 38). He adds that the "diagnostic red-on-brown Tlatilco ceramics and vessels bearing Olmec stylistic attributes occur at all times below Middle Formative levels at Chalcatzingo" (Grove 1974b: 55).

Tolstoy (1973: 13–14) appears to cling to the opinion that the Tlatilco-style burials are post-Olmec and "probably date from ca. 975 B.C. R.T. and from the decades that follow," although he feels that "some of the features . . . may be older in West Mexico . . . and, perhaps, even in Morelos." His temporal placement rests on his study of grave lots from Tlatilco and on the stratigraphy reported for Tlapacoya (Tolstoy and Paradis 1971: 23), which latter makes "Olmec culture in the Basin" of Mexico the earliest occurrence of ceramics in the series. However, Niederberger's (1976: 249, 278–79) recent work at Zohapilco establishes a long sequence extending from preceramic times (5500–3500 B.C.) through "incipient ceramic" to Zacatenco (800–400 B.C.) and thus demonstrates that the Tolstoy-Paradis stratigraphy is incomplete at the lower end. As for Grennes-Ravitz (1974: 102), he dates his La Manuela subphase, with its Tlatilco-style contents, at 1400–1200 B.C.

One of the chronic difficulties in placing offerings from the Tlatilco graves chronologically has been the inability to match the furniture with occupational debris. Inasmuch as the Tolstoy-Paradis sequence lacks the early end, it is not adequate for such equation. The Niederberger stratigraphy, just mentioned, has various elements of the Tlatilco style concentrated in the Manantial phase, 1000–800 B.C. (Tolstoy 1973: 14), and this material overlies remains of the Olmec-affiliated Ayotla phase. Ostensibly, then, the Tlatilco-style elements appear the more recent.

It turns out, however, that preceding the Ayotla phase, Mrs. Niederberger has isolated a "complex" that she calls Nevada (1400–1250 B.C.). She feels it is meagerly represented and should be strengthened. It, as well as Manantial, has elements reminiscent of the Tlatilco style: a bottle with tubular neck, but short; narrow gadrooning; and a red on brown color scheme (Niederberger 1976: 257).

In conversation, Mrs. Niederberger (personal communication, 1974) summarizes the situation as follows. There is an "early" occurrence of traits reminiscent of the Tlatilco style in Nevada context (1400–1250 B.C.). Next comes the phase called Ayotla (1250–1000 B.C.), during which these elements are eclipsed by a surge of Olmec-style products. Then these latter wane in the Manantial phase (1000–800 B.C.), and ceramic traits suggesting the Tlatilco style again are evident.

The above summary leaves us (1) with a Tlatilco style whose composition is unstable; (2) with major discrepancies in the reporting of its grave-lot associations; and (3) with differences of opinion concerning chronological position. Consideration of the latter aspect will be resumed below, following a discussion of resemblances between Capacha and the Tlatilco style.

Ceramics

In both Capacha and the Tlatilco style the preponderance of somber black-brown-gray monochrome is impressive. Nothing is known of nonfunerary Capacha products, but it has been suggested that the dark wares of Tlatilco may be "an index of prestige or wealth" (Tolstoy and Paradis 1971: 14).

The Tlatilco style has no bule as such, although it has cinctured vessels (Piña Chan 1958: Fig. 41 *j*; Grove 1970: Fig. 5 *f*) whose restricted orifice and exaggerated body constriction contrast with the open mouth and modest indentation of the Capacha bule. It has been suggested that the "belted bottle" of the Tlatilco style might have arisen as a cross between a Capacha-like bule and a non-Capacha tubular-neck bottle (Kelly 1972: 30). In this connection, the presence of "narrow-necked bottles" as an ingredient in the Ojochi and Bajío phases at San Lorenzo might be pertinent (Coe 1970: 22–25),[29] as is the presence of the bottle in Honduras (see this chapter: *Honduras,* below).

Although no stirrup pots have been found in the Opeño phase, this vessel form is a significant point of correspondence between Capacha and the Tlatilco style. However, close inspection indicates differences. The body of the Tlatilco specimens tends to be bell shaped, with a sharp wall-base angle (Porter 1953: Pl. 10 *e, f*; Piña Chan 1958: Figs. 43 *m,* 44 *n*; Grove 1970: Fig. 5 *e*); the orifice is restricted and not in the form of a small pot (Fig. 25 *b–d*).[30] Three-tube Capacha variants have no counterpart at the Tlatilco site, and the "jointed-tube" profile (Figs. 24 *b, c;* 25 *b–d*) is more suggestive of the Ecuadorean Machalilla phase (Meggers, Evans, and Estrada 1965: Pls. 155 *a,* 156 *m*) than of the Tlatilco style. Finally, all known Capacha stirrup pots are monochrome, in contrast to those of the Tlatilco style, which may be painted, some in resist.

Other vessel forms are less specialized, but there are general resemblances in cántaros, ollas, incurved bowls, and pinched-rim bowls (Piña Chan 1958: lám. 21, right, penultimate row). Capacha bases are almost invariably rounded, and Niederberger (1976: 14) notes that Tlatilco-site specimens figured by Piña Chan (1958) include relatively few flat-bottomed vessels. Brush (1969: 102) calls

attention to the same feature for Zacatenco, Ticomán, and El Arbolillo, in the Valley of Mexico, as well as for "the entire Preclassic span" of coastal Guerrero.

In decoration, broad-line incision is common to Capacha and the Tlatilco style and sometimes is combined with zoned punctation (cf. Fig. 22 *b–c,* and Piña Chan 1958: Figs. 37 *j,* 43 *e*). Partial zoning, with punctate fill, is common in Capacha monochrome, less so in the Tlatilco style. For the Manantial phase at Zohapilco—which contains certain elements of the Tlatilco style—there is mention of a garland (*festón*) element in decoration (Niederberger 1976: 268; cf. Appendix II: *Capacha Monochrome: Decoration*).

One specific resemblance in ornament involves definition of the design field by incised vertical lines, against which the base of an isosceles triangle is laid; opposed repetition leaves a residual hourglass zone (Fig. 30 *b, bb, d;* cf. Fig. 30 *h,* also Piña Chan 1958: Fig. 45 *j*). This zone and/or the flanking areas may be filled with punches or gouges (Fig. 30 *b, bb, d;* cf. Fig. 30 *g,* also Piña Chan 1958: Fig. 39 *h*). One specimen from the Tlatilco site (Porter 1953: Pl. 12 *b;* Piña Chan 1958: Fig. 37 *j*) has half an hourglass, with punctate filler, as the chief design. If the vertical dividing lines were eliminated, there would be some resemblance between the hourglass and the Capacha sunburst (see Appendix II: *Capacha Monochrome: Decoration*) and perhaps even between it and thè so-called net or lattice decoration, which is essentially a continuous, all-over version (Figs. 24 *a,* 26 *a;* Piña Chan 1958: lám. 20 right, top, shows an incomplete form). In a few instances, similarity in design between Tlatilco-site and Capacha-phase specimens is striking (Fig. 17 *h;* cf. Fig. 30 *g* and Piña Chan 1958: Fig. 39 *h*).

Rcd paint on the natural brown base ware, sometimes zoned with incision, occurs in both Capacha and Tlatilco-style pottery (Figs. 28 *c–f,* 30 *a–d;* Piña Chan 1958: 45, 46, Fig. 15). Resist painting is absent in Capacha and possibly present in Opeño, although handsome examples are known from the Morelos manifestation of the Tlatilco style (Grove 1970: 68, Fig. 5 *a, b*).

Figurines

Content of the Tlatilco style from the site of Tlatilco and from outlying areas is not identical, and there are differences with respect to the distribution and frequency of associated figurine types. Grove (1970: 68) comments that "K-variety figurines in the Pantheon mound [Morelos] burials predominate over D-1 and D-2 figurines in the sample, the apparent reverse of the situation at Tlatilco."

The few hollow Capacha figurines (Figs. 32 *b,* 34) do not conform in all respects to the usual Mesoamerican classification, but Profa. Rosa María Reyna, who has inspected them, feels they are basically K in type. I see comparatively little resemblance between the Capacha specimens and those illustrated from Tlatilco proper, although stance is reminiscent of a spouted model (Grove 1970: Fig. 6 *d*) from the Pantheon mound, and eye treatment and gopherlike mouth recall several figurines attributed to the same mound (Grove 1970: Fig. 5 *j, l, m*). Whatever it may mean, as far as figurines are concerned, Capacha shows stronger resemblance to Tlatilco-style examples from Morelos than to cognate specimens from the site of Tlatilco, and a closer similarity to Morelos products than to those of the Opeño phase.

Miscellaneous Manufactures

The inventory of Capacha miscellaneous products is so scanty that comparison with other areas is difficult. Although the Tlatilco site is richer by far, it is likely that a good deal of nonceramic material has been excavated since Lorenzo's report of 1965. Moreover, in the absence of published data concerning grave lots, little is known concerning association of most of the objects.

Considerable obsidian seems to have come from the Tlatilco site, but its cultural context is unclear. In Morelos, "blades" appear in Late Nexpa/San Pablo B (1050–900 B.C., to which both Olmec and Tlatilco-style elements are attributed; in contrast, chips occur in earlier association (Grove 1974a: 113; 1974b: 38, 48). It should be significant that in Capacha (Appendix III: *Obsidian*), in "early" context in Morelos, and at "Early Formative lowland [Gulf coast] sites" (Coe 1970: 22), obsidian is found as flakes or chips, not as prismatic blades.

One small stone object (Fig. 44 *b*) from a Capacha cemetery—as a guess, it might be an amulet—seems very close to the artifact that Lorenzo calls a polisher for pottery (1965: Fig. 39 1:1; cf. Kan, Meighan, and Nicholson 1970: no. 53, left). As with several traits common to Capacha and Opeño, the supposed amulets may be so general in the Mesoamerican Preclassic that the Capacha-Tlatilco correspondence is not meaningful. At least, such stones tend to substantiate an "early" temporal position for Capacha.

Comments

Despite the fluctuating composition of the Tlatilco style, several traits commonly associated with it show resemblances to Capacha. It is difficult to evaluate such similarities. The chronologies of Grove (personal communication, 29 September 1974) and Tolstoy (1973: 13–14) place the Tlatilco style almost half a millennium more recent than our one carbon date for Capacha.[31] This seems an excesive disparity, even if one believes diffusion to be a slow matter.

There may be an alternative. Owing to the apparently wide temporal span of certain Tlatilco style traits, it seems likely that with systematic analysis that provocative "style" might be broken into several components, the earliest of which might be roughly contemporaneous with Capacha and Opeño.

Honduras

Honduras lies on the fringes of Mesoamerica and only partly within it. Nevertheless, from this marginal area there are reports of stirrup pots, zoned and excised decoration, and other ceramic traits that sound familiar (Porter 1953: 89; Weaver 1972: 88); in fact, resemblances to Tlatilco are said to be so marked as to suggest same "kind of direct contact." Lowe (Green and Lowe 1967: 61–62, 64, 70–71) emphasizes the potential significance of Honduras to an understanding of the Mesoamerican Preclassic, pointing out the apparent contemporaneity of Yarumela

I with the earliest Mesoamerican products and noting, moreover, a suggestion of Valdivian resemblances in some materials from Honduras (cf. Baudez in Heizer and Graham 1971: 80).

Important ceramic material, most of it obviously, "early," has been reported from caves near Cuyamel, in the vicinity of Trujillo, Honduras (Reyes Mazzoni and Véliz 1974; Healy 1974). Unfortunately, the specimens can be placed only on stylistic grounds. Flat rather than rounded bases are characteristic. There are bowls, at least one incurved and one flaring; waisted vessels, neckless, which literally resemble bottle gourds; two bottles with cincture; and single-spout bottles with tall, narrow neck, as well as one with a tapering neck (compare Fig. 10 *d* with Reyes Mazzoni and Véliz 1974: Fig. 3, right). There is no bule illustrated, but one stirrup pot is shown by Porter (1953: Pl. 14); human and animal effigies are striking. Decoration includes red on brown bichrome, as well as fluting and gadrooning, but there seems to be no broadline incision combined with punctation.

At present, the relevance of Honduras is unclear. Both reports cited above (Reyes Mazzoni and Véliz 1974; Healy 1974) mention evident relationships to southern Mesoamerica and to Tlatilco. One wonders if Honduras may not have had a position on the Caribbean analogous to that suggested for Colima on the Pacific, during Capacha times (Kelly 1974: 209). It seems possible that maritime traffic along the Central American coast may have united Honduras, via a series of way stations with northwest South America; and from Honduras there was evidently communication with southern Mesoamerica proper. Published evidence indicates that Putun (Chontal Maya) canoe traffic traversed the local streams and circled the entire coast of Yucatán long before the Spanish Conquest (Thompson 1970: 7); Thompson (1970: 126–33) and Healy (1974: 442) summarize other references to coastal trade in the area. Conceivably, such traffic may have been established in remote times, as seems to have been the case on the Pacific coast—witness Capacha and its apparent ties with northwest South America.

Accordingly, there is a possibility that cultural influences from northwest South America worked northwest along both coasts, to enclose Mesoamerica in the arms of a skewed "V" with a northwest-southeast axis, across whose open or northern end South American culture traits moved inland from both sides to unite in central Mexico (at Tlatilco, for example). It may turn out that Honduras was a critical point on the eastern arm of this hypothetical line of communication, where influences from Mesoamerica (possibly the flat-base bowl and the tecomate, for example) and from South America (the single-spout bottle and the stirrup vessel) converged.

NORTHWEST SOUTH AMERICA

Resemblances

Capacha shows stronger similarities to northwest South America than to Mesoamerica, the exceptions being, of course, Opeño and the Tlatilco style. Although physically the two latter are situated within Mesoamerica, they have, as does Capacha, a perceptible cultural ingredient that links them to northwest South America.

A number of photographs of Capacha pottery were sent to Dr. Betty Meggers and Dr. Clifford Evans, who recognized several generalized features shared with northwest South America, including the small open-mouthed pot and "the broad incision, erratic form of the punctations, and polished surfaces"; they added that zoned punctation is diagnostic of a late form of Valdivia ceramics and characteristic also of Kotosh (personal communication, 3 October 1970). The impression of Drs. Meggers and Evans was reinforced in September of 1974, when they viewed the Capacha collection. Both they and Dr. Donald Collier, who saw the same material a few days later, felt that generalized resemblances to the early ceramics of northwest South America were discernible. Moreover, Dr. Gerardo Reichel-Dolmatoff, who saw a number of Capacha photographs, felt there was "something vaguely familiar" about the material but could not place it definitively; he added that "punctate-incised decoration . . . is fairly frequent in lowland Colombia, but it is not a time-marker nor is it diagnostic in any way" (personal communication, 7 November 1974).

A comparison with northwest South America obviously starts with Valdivia and Machalilla, of the Ecuadorean coast. Inasmuch as Valdivia is much older than Capacha, specific resemblances are scarcely to be expected. Yet in form, both have the incurved bowl, the small, open-mouthed pot mentioned above, and the rounded base, which latter is fundamental and enduring in Colima ceramics. The bewildering range of Valdivia decorative styles is wanting in Capacha, but there is a suggestive correspondence in the net or lattice arrangement (cf. Figs. 24 *a,* 26 *a;* and Meggers, Evans, and Estrada 1965: Pl. 77 *a–f*), although in Capacha the "hub" is a depression rather than a nodule. As far as is known at present, Capacha has no case of rocker or shell stamping, and its chief resemblance to Valdivia in decoration is broad-line incision, combined with punctation of various kinds. Finger grooving is another trait in common.

Valdivia is followed by Machalilla. Before inspecting possible ceramic correspondences between the latter and Capacha, mention may be made of a similarity in the type of cranial deformation known as tabula erecta. Interestingly enough, the same type of deformation occurs not only in Machalilla and Capacha association but is reported as well for El Opeño and the Tlatilco site (see conclusion of Appendix V).

Machalilla, like Valdivia and Capacha, has the rounded base, the incurved bowl, and the small wide-mouthed pot. Machalilla "finger punched" (Meggers, Evans, and Estrada 1965: Fig. 79) approximates the pinched-rim bowl (Fig. 12 *f, i*) of Capacha (and of several other "early" phases in Mexico). From no part of South America is a bule form reported archaeologically, but a single sherd (not illustrated) of "Machalilla Red incised," described as having "a complex wall profile consisting of two circumferential concave zones separated by a pronounced circumferential ridge (Paulsen and McDougle 1974: 3), evidently has some sort of double cincture.

The one spectacular parallel between Capacha and western South America is the stirrup-spout vessel. This correspondence has not passed unnoticed—Coe (1963: 103) thinks that the form is not "natural," and he therefore suggests "a single point of origin as the most reasonable hypothesis to account for its distribution in the Western Hemisphere." In considering resemblances, the special features of such Capacha pots should be borne in mind: the aperture in the form of a small vessel, sometimes a miniature; the comparatively generous orifice, in contrast to the narrow tubular spout of other areas; the preference for three, rather than two, connecting tubes; and the "elbow" or "broken" profile of the tubes, irrespective of number. These characteristics are not invariable. In one example, the upper vessel is not sharply delineated (Fig. 25 *e*); in another, the lower part of the vessel is of doughnut form and is somewhat dwarfed by the upper body (Fig. 24 *c*).

Several summaries of stirrup-pot distribution have been published (Brainerd 1949; Phillips, Ford, and Griffin 1951: 171–72, 205–6; Kelly 1974: 208). In addition, a few effigy stirrup vessels are attributed to Nayarit, although nothing is known concerning their association, and the same is true of the small trifids from Sinaloa, whose tube form is remarkably similar to that of the Capacha vessels (see Chapter 3: *Distribution: Nayarit and Sinaloa;* Chapter 4: *Nayarit,* above). Likewise from Sinaloa come polished red sherds, apparently stirrup-pot fragments (Kelly 1945a: 105–6, Fig. 57 *f*).

Temporal discontinuity is marked. Some vessels, such as those of Capacha, of the Tlatilco style, and of Chupícuaro, are "early" and are scattered through the equivalent of the Mesoamerican Preclassic. Several isolated human-effigy specimens may belong to that same general time level: two attributed to Nayarit (one of them reported in Kelly 1972: 28) and another thought to be from Colima (Corona Núñez 1960a: [Fig. 9]). In contrast, a much later clustering of non-effigy specimens occurs in the lake area of Michoacán, at Coyuca (Guerrero), and at Culiacán (Sinaloa).

Capacha stirrup vessels tend to be heavy and massive and lack the delicacy and elegance of the late forms—for example, those of highland Michoacán. This massive aspect makes for a general resemblance between Capacha and Machalilla specimens, although the latter are described as "squatty" (Meggers, Evans, and Estrada 1965: 139), while the Capacha two-tube pot (Figs. 13 *d;* 25 *e*) is rather tall and ungainly, the three-tube variants, less so. A few Machalilla examples have a faint hint of the "elbow" (Meggers, Evans, and Estrada 1965: Pl. 156 *c, m;* Fig. 88, bottom row) and of a vessel-shaped aperture (Meggers, Evans, and Estrada 1965: Pl. 156 *f*). I am indebted to Dr. Allison Paulsen for calling attention to a Machalilla stirrup pot, whose spout is a miniature vessel in the form of a small incurved bowl with sharp midbelly carination (Estrada 1957: 47, Fig. 19 *b,* central sketch; cf. spout of somewhat similar form mentioned in Chapter 3: *Distribution: Nayarit and Sinaloa*).

During perhaps a millennium, an extraordinary assortment of stirrup-spout vessels was made on the coast and in the highland of northwest South America. Apart from Machalilla, this remarkable shape is reported for Chavín and its coastal counterpart (Cupisnique), Kotosh-Kotosh and Kotosh-Chavín, Recuay, Salinar, Mochica, Gallinazo, Chimú, and others. Some Chavín specimens evidently are undecorated, but they seldom are illustrated. Outlines of two (Lumbreras 1971: Fig. 6) indicate a squat form that sits heavily on a flat base. Proportions are very non-Capacha, and most published Chavín specimens are so lavishly ornamented as to be remote from Capacha.

The same is true for most other stirrup pots illustrated from northwest South America, but two similarities—probably not significant because of timing—may be noted.

1. One is a black-ware vessel whose aperture has the form of a small tumbler with straight flaring sides, not unlike the upper part of a miniature pot attributed to the Capacha phase (Meighan 1974: Fig. 3, right). The black-ware specimen is ascribed to the "Middle period" on the north coast of Peru (Bennett 1946: Pl. 41 *e*), hence is too recent to relate to Capacha.

2. The other resemblance concerns trifid tubes. Few examples are recorded for Nuclear America, except in presumed Capacha association (see Appendix II: *Capacha Monochrome: Form*), although Brainerd (1949: 7) mentions an "aberrant form from Jalisco, with an extra spout from the body." Far to the north, a lone case is reported from Arkansas (Phillips, Ford, and Griffin 1951: Fig. 106 *e*).[32] This shape is not illustrated for Machalilla, hence casual mention of seven trifid-spout vessels in the Recuay style of the highlands of Peru (Bennett 1944: 102–3, Fig. 32 *L*) suggests striking correspondence with Capacha. The tubes are described as "meeting in an open bowl-like mouth" (the latter not very evident in the sketch). Moreover, trifid tubes are said to "occur on ring-shape containers and on other styles." This again suggests Capacha, for one triple-tube vessel, almost certainly of that phase, has a doughnut base (Fig. 24 *c*). Indeed, unless there are changes in chronology, the Capacha specimen may be the earliest ring-shape vessel known at present from Nuclear America (cf. Parsons 1963: Fig. 2; Dixon 1964: 458).

The Capacha-Recuay combination of ring base and trifid tubes seems necessarily significant, although other Recuay specimens illustrated by Bennett (1944) are not reminiscent of Capacha, and the comparatively recent chronological position of Recuay (personal communication from Dr. John Rowe, 11 November 1974) makes meaningful relationship to Capacha unlikely. One wonders whether there could have been a common source, with late survival in Recuay.

Any discussion of relationships between west Mexico and South America tends to focus on shaft tombs, and most of the forms illustrated by Long (1967) for northwest South America can be matched in Colima. There is a strong possibility that the tomb occurs in Capacha times, for grave goods assuredly of that phase have been removed from shaft tombs, according to independent accounts from several Colima informants (see Chapter 3: *Interments*). The Opeño variety of stepped-shaft tomb (Noguera [1942]; Oliveros 1974) dates from 1500 B.C. (uncorrected date), and any Capacha tomb should be roughly synchronous. A comparatively early use of the shaft tomb in Colima is relatively certain, even if post-Capacha.

Sherds collected from the edges of a rifled tomb in east Colima, and provisionally classed as Zapote rose, yielded a radiocarbon date of 1130 B.C. (GX-1785). (This date is not included in Table 1, because it refers to east Colima.)

Information concerning tomb chronology in northwest South America is scanty. Willey (1971: 277) places tomb interment within his Formative, which has a generous time span from 1500 to 500 B.C. Weaver (1972: 287) suggests a date of 500 B.C. for Colombian tombs, but Dr. Gerardo Reichel-Dolmatoff notes that all dates he has for (presumably Colombian) shaft tombs are "A.D. and therefore much too late" (personal communication, 7 November 1974). For Ecuador, Evans and Meggers (1966: 259) give a date of 500 A.D.; Dr. Donald Collier thinks such tombs are not earlier than the Chorrera phase, which he dates from 1100 to 300 B.C., and he adds that they may be confined to the latter half of the phase (personal communication, 31 December 1974). As matters now stand, the Opeño date of 1500 B.C. appears to give Mexico clear temporal priority.

The Chronological Quandary

Manifestly the chronological situation is basic to any attempt to view Capacha–South American resemblances in perspective. Capacha dating leaves a good deal to be desired, but its one radiocarbon date of 1450 B.C. is mightily bolstered by the Opeño date of 1500 B.C.

It turns out, however, that the original guess date of 2000 B.C. for the start of Machalilla is a matter of some controversy. Several investigators feel it should be moved upward to a still conjectural 1500, 1400, 1300, or even 1100 B.C., with a correspondingly more recent terminal date, placed variously from 1100 to 300 B.C. (Lanning 1967: 85; 1968: 47, 49; Paulsen and McDougle 1974; personal communication from Donald Collier, 31 December 1974).

Such revision would make Machalilla more or less contemporaneous with Capacha and Opeño, and perhaps with some aspects of the Tlatilco style, providing these latter are extended backward in time (see this chapter, *Tlatilco Style: Comments*).[33] Inasmuch as the stirrup pot is not reported for early Machalilla (Meggers, Evans, and Estrada 1965: Fig. 90, Table G), Capacha vessels of that form may be on the same time level as those of Machalilla, if not actually earlier.

Even as Machalilla is being updated, Chavín apparently is being pushed backward in time so that, although the two are not considered quite coeval, they are no longer widely separated temporally. In fact, Paulsen and McDougle (1974: 13–14) suggest that Machalilla may have derived its stirrup pot and associated single-spout bottle through trade in "early Chavín or even in proto-Chavín times," adding that Machalilla products may represent "rustic simplifications of the elaborately decorated stirrup-spouts and spouted bottles that characterize Chavín pottery styles."

The upshot of this chronological scramble is that (1) Machalilla has lost its temporal priority relative to Capacha; and (2) if the revision downward of the dating of certain Tlatilco-style traits proves acceptable, Capacha no longer predates the latter by half a millennium. It may be added that the stirrup spout is not one of the Tlatilco-style elements to appear in the Niederberger sequence at Zohapilco (this chapter, *"Tlatilco Style": Chronology*).

A few years ago, when the initial notes concerning Capacha were prepared (Kelly 1972, 1974), I counted on a start of 2000 B.C. for Machalilla and one of 975 B.C. for the Tlatilco style. The suggested interpretation was simple and apparently satisfactory in terms of the then current chronology. Machalilla preceded Capacha by 500 years, hence Ecuador seemed a logical source of Capacha stirrup vessels and possibly other traits that tied with South America. Furthermore, Capacha was 500 years earlier than the guess date for the beginning of the Tlatilco style and thus might have been the source of several shared elements, such as a preference for dark monochromes, pottery decoration with broad-line incision and punctation, and stirrup-mouth vessels. Finally, Capacha, on the Pacific coast, enjoyed a strategic position with respect to maritime contact.

With recent shifts in dating, Capacha's comfortable temporal bolsters are collapsing, and all the cultures in question seem to be more or less coeval. As far as northwest South America is concerned, the sheer preponderance of its early occupation—with deep deposits for Valdivia, although not for Machalilla—and its cultural diversity seem to rule out the likelihood of major north-to-south coastal influences from Capacha. The latter, with its few small cemeteries and with no occupational sites yet identified, is not likely to have inspired great cultural developments in the southern continent. Yet there remains the fact that west Mexican tombs seem appreciably earlier than those of northwest South America.

As for Capacha and the Tlatilco style, the cultural ballast is markedly heavier on the part of the latter, and sites with related ceramics are turning out to be more numerous and more widespread in central Mexico than was realized a few years ago (Grove 1971: 24). It was not difficult to overlook the uneven weighting when Capacha-Opeño occurrences dated half a millennium earlier than the Tlatilco style.

Even so, the possible explanations are limited, and that suggested some time ago (Kelly 1972, 1974) still may be —with some modification—the most economical one, provided it does not do violence to the chronology. The so-called Tlatilco style is found in landlocked country, with no possibility of direct sea contact with northwest South America. Capacha, on the Pacific littoral, seems a likely way station for cultural influences that moved up the Pacific from northwest South America, extended northwest along the coast, at least to Sinaloa, and pushed inland from the coastal base (see Chapter 3: *Distribution*). Honduras remains an enigma, but it also might have been an intermediate point on the Caribbean for South American cultural impulses, which spread therefrom to central Mexico.

Theoretically, were it not for the apparent discrepancy in time, Mexico-Morelos might have been the source of both Opeño and Capacha developments—an explanation that could account, at least, for a certain resemblance between Capacha figurines and some examples from Morelos. But a central Mexican source would not explain for Capacha the weakness or absence of the narrow-necked bottle, the apparently local development of the bule, and the absence of resist painting.

The situation must be examined anew once more information is available and the chronological dust has settled.

SUMMARY

To recapitulate, Capacha is Capacha. It is not Mesoamerican, yet not quite South American, although perceptible ingredients link it with the northwest part of South America. To a considerable extent, Capacha confirms the prediction of Ford (1969: 166), who foresaw that "the earliest pottery of coastal Mexico should show more resemblance to Puerto Hormiga, Machalilla, or Valdivia" than to the Tehuacán tradition. Now that attention has been called to Capacha, the chances are that similar material will crop up in many parts of west Mexico. If the suggestion of maritime contact is correct, Capacha may represent no more than one of several landfalls along the Pacific coast.

The statement of Drs. Meggers and Evans (personal communication, 3 October 1970), covers South American resemblances neatly. They feel that Capacha cannot quite be duplicated in the South American collections, "but all the elements are present," and that Capacha relates "to the general content of the early Formative of Ecuador, and its unique combination of elements from this early tradition follows the precedent of previously discovered complexes like Puerto Hormiga, Barrancoid, Kotosh, Barlovento, Monagrillo," and others.

By no means is Capacha a copy of anything known from northwest South America—or elsewhere, as a matter of fact. It has its own special features, such as the highly distinctive bule with its equally distinctive sunburst ornament in incision and punctation. Its peculiar stirrup pots, usually trifid, and with "elbowed" tubes, are unique. Certain absences must be significant but are difficult to explain. One is the paucity, if not the actual lack, of the narrow-necked, single-spout bottle (see Appendix II: *Capacha Monochrome: Form*); others, the absence of shell and rocker stamping. Capacha figurines are totally distinct from presumably contemporary South American products and bear little resemblance to several singularly handsome Opeño specimens (Oliveros 1974: Figs. 19, 20), although some are reminiscent of certain Tlatilco-style figures from Morelos. In addition, the variety and excellence of Capacha ground stone artifacts merit mention.

The idea of basic relationships between Mesoamerica and South America is of long standing, and so also is that of some sort of contact between west Mexico[34] and western South America. The Capacha data simply add more weight to the affirmative side of the discussion and on a time level earlier than other evidence has suggested. A long bibliography concerning the presumed contact has accumulated, but here no effort has been made to cite all such studies; only those that seem particularly relevant to the Capacha phase are mentioned below.

Several investigators have felt that the Tlatilco style is not an integral part of Mesoamerican culture. Lowe (Green and Lowe 1967: 70) suggested communication with "Peru on the Chavín horizon" and believed the Tlatilco style represented an intrusion resulting from Pacific coast landfalls, possibly in Guerrero. Tolstoy and Paradis (1970: 349; 1971: 18) agreed that this same "style" had basic South American ties attributable to influences that reached central Mexico via the Pacific. Subsequently, Tolstoy (1971: 26) suggested that the Tlatilco style should be situated "within West Mexico, rather than within Mesoamerica proper." The preceding comments concerning the non-Mesoamerican affiliations refer expressly to the Tlatilco style but by implication apply equally well to the Capacha and Opeño phases.

Tolstoy (1971: 26) felt that the situation might be viewed in terms of a "shifting boundary . . . between two major traditions (the Mesoamerican and the West Mexican) in the Central Highlands between 1200–950 B.C." With this dichotomic interpretation I am very much in agreement but think it may have to be modified to admit the possibility of dual infusion by maritime contact from northwest South America—one route via west Mexico and perhaps another via Honduras. Furthermore, the dating will have to be pushed back some centuries to accommodate Capacha and Opeño.

To summarize, although Tolstoy was not thinking specifically in terms of Capacha, this phase seems to me to epitomize what he called the "West Mexican tradition." Its resemblances to material from northwest South America are more marked than are those of more recent phases. Above all, with respect to contact along the Pacific, Capacha occupies a key position, concentrated as it is in Colima and adjacent Jalisco, with a probable extension northwest along the coast, at least to central Sinaloa (see Chapter 3: *Distribution*). Presumably this strip provided the footing for a thrust inland, to include Opeño, possibly the Tlatilco site, and various other central-Mexican localities that share the Tlatilco style. In addition, there is some chance that, more or less concurrently, cultural influences likewise from northwest South America moved up the Caribbean coast to Honduras, and thence to central Mexico (see this chapter, *Honduras*), thus closing what might be called a cultural circuit.

As a final note, the possibility of Capacha relationships outside the New World has not been overlooked. Because of the suggestion that the Valdivian culture of Ecuador stems from the ancient Jomon culture of Japan (Meggers, Evans, and Estrada 1965: 158–71), photographs of representative Capacha pottery were sent to Dr. Wilhelm Solheim II and to Dr. Chiaki Kano. The former (personal communication, 24 September 1970) replied in words almost identical to those of Drs. Meggers and Evans quoted above, to the effect that Capacha cannot be duplicated in southeast Asia, "but all of the elements in form and decoration are present." In contrast, Dr. Kano (personal communication, February 1972) saw no specific resemblance between Capacha and the early "Jomon or Yayoi" wares of Japan.

Appendix I
CEMETERIES AND TESTS

1. ARROYO DE SAN ANTONIO

The Arroyo de San Antonio, Apulco, Tuxcacuesco, Jalisco (Fig. 1, no. *1*), was visited in 1942, at which time it was being destroyed by an excavation for making adobe bricks. The brief observations made then may be repeated:

Site at the mouth of the small arroyo which enters the Tapalpa at Apulco, in the bottom of the gorge. No surface evidences; sherds collected from an excavation made for adobe. Excavation apparently had disturbed one or more free burials; bone fragments noted a meter below the present surface, in the wall of the pit. No rubbish of occupation. In some respects the ceramic material is aberrant ... but it has definite ties with the Tuxcacuesco complex (Kelly 1949: 210).

The "aberrant" ceramic material includes (1) a heavy, incised bule, purchased in Apulco (Kelly 1949: Fig. 59 *c*). It also includes (2) a large, handsome cántaro restored on the basis of surface sherds; the vessel is unique, of zoned black and red, with the incisions filled with a white substance, perhaps lime (Kelly 1949: Pl. 14 *d*). With no counterpart in the present Capacha collection from Colima, this black and red vessel ties gratifyingly with the contemporaneous material from El Opeño, Michoacán (see Chapter 4). Lastly should be mentioned (3) a sherd with unusual incision (Kelly 1949: Fig. 59 *b*) and with a slight horizontal channel at the base of the neck, a feature characteristic of certain Capacha ceramics (Figs. 10 *d*, *f*, *g*; 11 *h*, *i*).

These three specimens establish the presence of the Capacha phase in the Apulco-Tuxcacuesco area, and recently a small but impressive collection of Capacha-phase vessels has been reported from the vicinity of Apulco (Meighan and Greengo 1974).[20] My early suggestion of ties with the Tuxcacuesco phase is easily explained. The latter is the local equivalent of Colima's Ortices phase, and it was mentioned because a "restorable bowl of Tuxcacuesco red ware, which presumably accompanied one of the San Antonio burials, suggests that these incised vessels may equate with that phase, but the association is not certain" (Kelly 1949: 84).

It now is possible to attribute the "aberrant" incised wares from the Arroyo de San Antonio to the Capacha phase and, furthermore, to say definitely that Capacha is earlier than Ortices and its Tuxcacuesco equivalent. Evidently, there were several burials adjacent to the arroyo— at least one, of the Tuxcacuesco phase; the other(s), Capacha.

2. LA CAÑADA

Description and Tests

The Capacha cemetery at La Cañada (Fig. 1, no. 2) is situated 3–4 km west of modern Comala, within the municipal unit of Colima, fairly low on the flank of the Volcano. To the northeast is the impressive Volcán de Fuego; to the northwest, the great calabash-shaped Cerro Grande, across the Armería and across the state boundary, in Jalisco. This is beautiful open country, nowadays used chiefly for planting maize every alternate year and for raising cattle. A small stream has been diverted for limited local irrigation.

However, the Capacha cemetery is not in open country, but in an old planting of tamarind, lima, and coffee (Fig. 2 *a*), about 100 m from the southeast edge of the declivity that gives La Cañada its name. This is a depression perhaps 20 m deep and 200 m wide, through which a bare trickle of water runs. Vegetation is lush, and great *Ficus* trees are interspersed with coffee, mango, and citrus.

Apparently this cleft is what remains of an old watercourse tributary to the Arroyo Seco; the latter—now enormous, stony, and stark—must at one time have been an important branch of the Río Armería. Here, as in many parts of Colima, the opinion of a geologist would be enlightening. It may be assumed, I think, that during Capacha times the La Cañada cleft was the course of a modest stream. There is indication that during the era of the shaft tombs—perhaps Ortices, perhaps Comala—La Cañada was already rather deeply cut. At least, on a narrow bench well down the slope of the supposed old channel there is a very large deliberately pitted boulder, similar to those popularly considered "maps" recording the location of shaft tombs. Seldom is specific correspondence demonstrable, but such pitted stones are found in the vicinity of cemeteries with tombs and are taken seriously by moneros.

The La Cañada cemetery was discovered by chance in the spring of 1970, when several moneros stopped to eat in the shade of its trees. Idly, as they rested, they dug small pits; for any dedicated monero, such test pitting is almost automatic. To their surprise—for there were no surface indications—they came upon the mouth of a shaft tomb. They cleared it, finding in the chamber a single dog-eared vessel of Comala red ware, of poor quality and incomplete.

Seldom do tombs occur singly, so the moneros started to dig vigorously—without, needless to say, the knowledge of the owners of the property. No additional tombs were found, but the moneros seem to have opened about 20 pits, all told, some outside the limits of the cemetery.

a

b

Fig. 3. Interments at La Parranda and Quintero.
a, La Parranda, Site A, Burial 2 (see p. 43). *b,* Quintero, Burial 8; in the upper left, an iguana hole; no bone accompanied this presumed interment (see p. 51).

On inspection three weeks later, each pit appeared about 1 m in diameter on the surface, but as it approached subsoil, where interments occur, it was undercut to such an extent that the actual diameter might be four or even five m. Inasmuch as this particular cemetery does not exceed 20 by 30 m, the undisturbed area was comparatively small. Most of the monero tests seem to have produced exclusively Capacha furniture, although several free burials on the western fringes were accompanied by later material.

The bulk of the offerings uncovered by the moneros either were found broken or were destroyed by ruthless digging; many fragments were abandoned on the surface, and on my visit I collected a skull (Appendix V, no. 9638) and a large quantity of sherds. The very few entire or nearly entire pieces were retained by the moneros, and I was able to purchase most of them because dealer interest had been negligible. The monero excavations were on a larger scale and more productive than mine, but the combined data shed welcome new light on the Capacha phase (see *Comments* below).

At La Cañada, as at most other Capacha cemeteries, excavation necessarily was a matter of fitting tests into the areas not churned by monero activity. Soil was extremely hard (*amasado,* or "kneaded," as the workmen put it) and contained a good many small bits of limestone. Probably the workmen were correct in their belief that a deliberate mixture of mud and small stones was spread as a protective layer over the interments, although it was not possible to trace the outlines of such covering.

I dug four tests at La Cañada. Two reached subsoil at a depth of 70 to 80 cm; they had few sherds and no burials. The two remaining tests produced interments.

Test 1

Originally 120 by 160 cm. Area immediately east of monero pit said to have yielded a double plate (Fig. 26 *e*), a small, dubiously classified vessel with rose wash and finger grooving (Figs. 28 *b*; 29 *a*); a small bowl of execrable quality (Fig. 12 *l*), as well as a bule and an "unfired" figurine; neither of the latter available for inspection.

In the west corner of our test, 50 cm below the surface, was a cluster of fragmentary long bones, not articulated. Without ceramic association, they were collected as a carbon sample (no. 229). At the same depth, roughly in the center of the test, we found a small Capacha-monochrome olla, of poor quality (Fig. 11 *h*), sitting upright. Small bits of bone were scattered throughout the test, indicating reuse of the cemetery; the olla just mentioned seems to be all that remained of a destroyed interment. Sherds few, mostly Capacha.

In the same west corner of the test, 80 cm below the surface, hence lower than the cluster of long bones mentioned above, we found fragments of leg bones; balance of the skeleton within range of adjacent monero pit; this bone not saved.

In the east corner of the pit, further leg bones (Burial 1) cleared, at a depth of 80–90 cm. They ran at an angle into wall of test, and the pit was extended east and south to clear the interment. Bones in extremely poor condition;

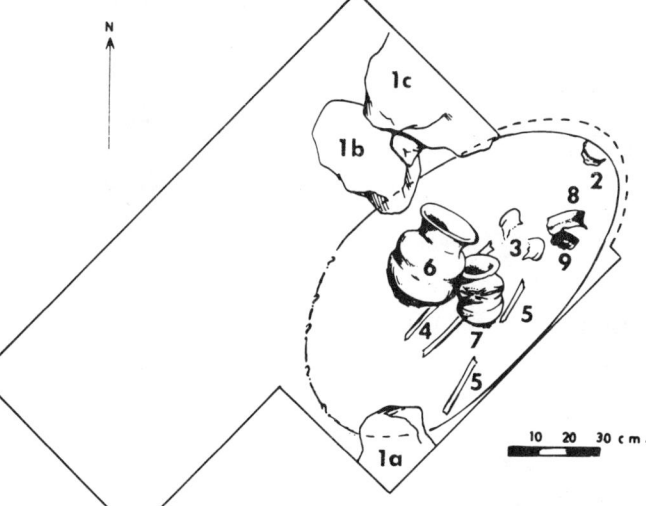

Fig. 4. Interments: La Cañada, Capacha-phase group Burial 2. *1 a-c,* large stones; *2-5,* skeletal material; *6-8,* potterly; *9,* fossilized limestone.

Partially overlying Burial 2 was Burial 1, which has been removed. In the line drawings in Figures 4 through 8, the top of the sketch is toward magnetic north. (See pp. 24, 91.)

Shown in the sketch are the remains of at least two persons in a shallow (15 cm deep), roughly oval pit, cut in subsoil that is 120–125 cm. below the surface. The pit is well defined, except on the southwest. To the northeast we undercut to trace the outline of the grave and to clear the mandible (*2*). The latter is directly against the edge of the pit and no cranium was found. Large stones, *1 a,* approximately 90 cm below the surface; *1 b,* 120–130 cm, *1 c,* 100–120 cm. Beneath *1 b* and *1 c* are other stones, not visible.

One skeleton was extended, with the head to the northeast; the position of the mandible (*2*) suggests a supine position, and fragments of pelvis (*3*) and long bones (*4*) align. A second interment, presumably earlier, is represented by tiny, scattered bits of

skull (not shown) in the vicinity of the bules, and fragments of two long bones (*5*).

Furniture at the knees of the more complete skeleton consists of one undecorated bule (*6;* Fig. 9 *h*) and one with characteristic Capacha grooved and punctate ornament (*7;* Figs. 15 *a;* 17 *b*). Possibly a relic of the furniture that accompanied the earlier and more fragmentary interment is the rim (*8;* otherwise not illustrated) of a large bule. It lies on top of a thin slab (*9;* otherwise not illustrated) of "fossilized limestone."

The photograph shows an early stage in clearing this burial; bules and bule rim have been exposed, but not the bone and not the outline of the subsoil pit.

none complete, but sufficient material to indicate the skeleton was extended, head to the east, face down; hands together at the back, with elbows bent. Apparently a small and fragile individual, but not an adolescent, for several teeth badly worn. A rough measure, in situ, suggested stature of 140–145 cm. No vertebrae; pelvis barely traceable; skull merely sufficient to establish position. Bone collected as carbon sample 230.

No furniture, hence phase allocation of Burial 1 uncertain, but it is suggestive that two interments at Quintero (Table 2, Burials 1, 6) were likewise prone, although they too lacked offerings that would permit phase allocation. Along north side of skeleton, at slightly higher level, two sizable stones aligned with body. Latter not contained in a prepared grave, but overlying, in part, a Capacha burial in a pit cut in subsoil (Burial 2; see Fig. 4). Skeletal material extremely fragmentary but included remains of two, possibly three, persons in a shallow, oval pit cut in subsoil. Bone collected as carbon sample 231.

Test 2

Four meters south of Test 1, another pit was opened and from it came Burials 3, 4, and 5 (Fig. 5).

A large bule (Burial 4; Fig. 5, no. *8*) was found in an oval subsoil pit. The test was extended to uncover, to the northeast and east, at higher level, a "pavement" of widely-spaced limestones. On top of one of them (Fig. 5, no. *1 e*) were sherds of two unclassified vessels, one Plain, the

other with red throat(?). The "pavement" clearly postdates Burial 4 of the Capacha phase, because pit of said burial underlies one of the paving stones.

An extension southwest uncovered two additional burials situated in another pit in subsoil (Fig. 5). Burial 5 was a secondary interment consisting of bone fragments of two adults (Appendix V, no. 10293 a, b) within a bule.

Comments

Ceramic material from La Cañada is overwhelmingly Capacha, and all offerings found in our two tests are of that phase. Nevertheless, later use of the site is evident. The Comala-phase tomb on the western boundary of the cemetery has already been mentioned, and, to judge from two vessels purchased from moneros, there may have been a few free burials of Ortices-Comala and of Colima-Armería affiliation. A third purchased vessel has been classified dubiously as a Capacha variant (not illustrated; Appendix IV-E-2, 9641 a). It is a medium-sized olla, with four equispaced maroon stripes pendent from the rim.

Likewise attributed to the La Cañada cemetery is a tripod bowl that moneros call a molcajete (grinding bowl), although its floor is not prepared for special use. Such pseudo-molcajetes probably belong to the Comala phase. The specimen is definitely not Capacha and does not appear in Appendix IV.

Three figurine fragments also are non-Capacha. One of

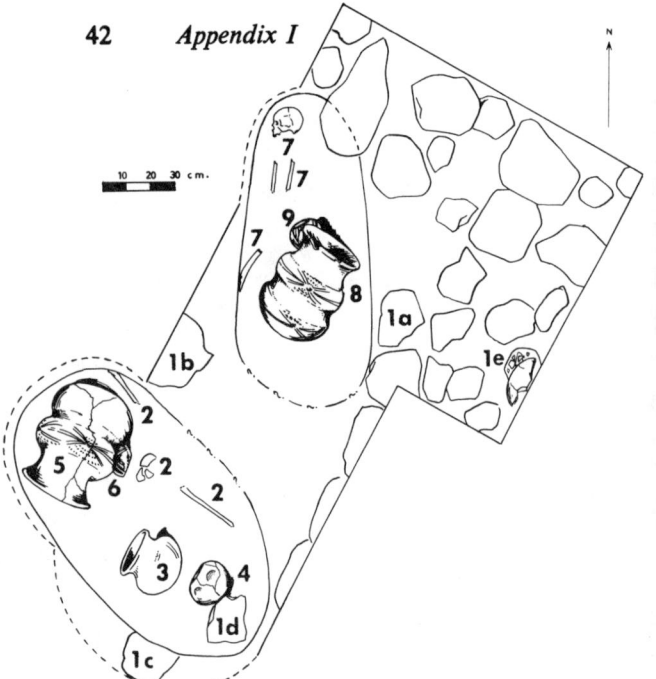

Fig. 5. Interments: La Canada, Capacha-phase Burials 3, 4, and 5. *1 a-e,* stone slabs; *2,* Burial 3; *3, 4, 8, 9,* pottery; *5,* Burial 5; *6,* unfired figurine; *7,* Burial 4. (See pp. 24, 27, 41–43, 91.) Two shallow, nearly-adjacent burial pits were cut in subsoil, both with the pit floor approximately 160 cm below the surface.

Burial 3 (*2* on sketch) is contained in the southwesterly pit. It consists of a few bits of skull and fragments of two long bones. Distribution suggests either more than one individual or major dislocation of a single interment, perhaps when the large bule burial *(5)* was introduced. Two vessels of Capacha monochrome are considered offerings associated with Burial 3, although they are somewhat removed. One *(3)* is an undecorated olla (Fig. 11 *g*); the other, an incurved bowl (*4,* Figs. 22 *e;* 23 *c*), with grooved and punctate decoration.

Burial 5 *(5)* is secondary and is contained within a large, decorated bule (otherwise not illustrated; Appendix IV–A–2, 9722) within the same pit as Burial 3. Bones are fragmentary but represent two adults (Appendix V, no. 10293 a, b). The bule rests directly on top of the crushed remains of an unfired figurine (*6,*

Fig. 32 *a*), perhaps a relic of the offerings with Burial 3.

Burial 4 *(7)* occupies the pit to the northeast, and this evidently predates a rough "pavement" of limestones. Only the stones associated with the presumably post-Capacha pavement are shown. These are not closely spaced; most are about 80 cm below the surface, but placement is extremely uneven. Slab *1 a* is deeper than the other stones; *1 b* is at an angle and extends from 110 to 130 cm below the surface; *1 c* and *d,* from 125 to 140 cm. On slab *1 e* were several sherds of a Plain pottery bowl and of another vessel with red rim; both are unidentified but are not of the Capacha phase.

Burial 4 consists of one individual, head to the north, face looking west. Roughly aligned long bones indicate an extended interment. Furniture consists of a large, decorated Capacha bule (*8,* Fig. 17 *e*), its rim resting on an incurved bowl *(9),* which is a red on brown variant of Capacha monochrome (Fig. 28 *f*). In the photograph the exposed bule *(8)* rests against the incurved bowl *(9),* but the pit in subsoil has not yet been traced and the one femur has just come into view.

these, purchased, is a typical Ortices-Tuxcacuesco-Comala specimen. Three other fragments, from general digging, are small and solid; one has a cusped foot. These will probably prove to belong somewhere in the Ortices-Comala range.

Except for monero discards on the surface, La Cañada produced few sherds. Only in Test 1 was an effort made to separate by levels. There, Capacha-phase fragments were comparatively more plentiful below 50 cm.

In the upper cut was an assortment ranging from Capacha (47 sherds) and a couple of dubious Ortices fragments (2) through Comala (16). The latter, as well as some sherds of Amoles wiped (11) may be explained by the Comala-phase tomb and possibly one or more free burials on the west fringes of the cemetery. Awaiting clarification is the comparatively large number of fragments of Red unclassified ware, badly incrusted (153 sherds). These may prove to relate to the Red unclassified found at La Parranda B, nearby; neither lot has been carefully studied as of 1975. In addition, the upper level of Test 2 produced a few sherds (11) vaguely reminiscent of the Colima-Armería phases, and the "pavement" adjacent to Burial 4

(Fig. 5, no. *7*) and partly overlying it also belongs to one of these sometimes indistinguishable phases, judging from fragments of two vessels resting on one of the stones (Fig. 5, no. *1 e*).

Significant information from La Cañada may be summarized thus:

1. Our only instance of a compound plate (Figs. 13 *b;* 26 *e*).

2. One certain occurrence of a Capacha monochrome painted variant, red on brown, not zoned by incision (Figs. 5, no. *9;* 28 *f*); it was found in direct contact with a Capacha bule. Another specimen of unzoned red on brown—purchased, hence without association—may possibly be a Capacha variant (Appendix IV-E-2, 9641 a).

3. Clear evidence of two vessels with zoned red or brown decoration (Fig. 30 *b, bb, c, cc*), one with punctation filler in combination. Both pieces were assembled from monero discards left on the surface and reputedly had been found in association with a cranium (Appendix V, no. 9638) mentioned above. It appears that these two painted and incised vessels can be attributed to the Capacha phase with reasonable confidence.

4. A small rose-slipped pot has decorative finger grooving (Figs. 28 *b*; 29 *a*), and similar decoration occurs on a cántaro from the Capacha cemetery proper (Fig. 26 *b*). However, paste and color are atypical, and the specimen is not certainly Capacha.

5. One of the two fragments with excised decoration (Fig. 31 *k*) from Colima comes from general digging at La Cañada; it may be considered a trade sherd.

6. Capacha-phase figurines are known particularly from La Cañada. As the moneros gave a pit-by-pit account of their endeavors, they referred frequently to *monos crudos* (unfired figures). The description is apt, and we collected several fragments that had been discarded on the surface (Figs. 32 *b*; 33 *b*, *c*, *e*). Fortunately, it was possible for the vendor to repurchase two entire specimens from La Cañada (Figs. 34 *a–aaa, b–bbbb*), which he had sold a couple of weeks earlier to a dealer.

A somewhat different figurine style, but likewise Capacha, was found beneath the bule that contained double Burial 5, a secondary interment. It must have been quite large, with a flat, solid head (Fig. 32 *a*) and hollow body; of the latter, no trace remained, and even the head is not fired. Still another figurine, unfired (Fig. 33 *a, aa*) was found within the bule that contained the secondary burial just mentioned (Fig. 5, no. *5*).

7. The only case of urn burial known thus far for the Capacha phase is that of the large bule noted above (Burial 5; Fig. 5, no. *5*). It contained the bones of two fragmentary adult skeletons (Appendix V, no. 10293 a, b).

3. LA PARRANDA, SITE A

Description and Tests

La Parranda, site A, Comala (Fig. 1, no. *3*), is about 3 km southeast of La Cañada, in a direct line over fields and part of an intervening property named Guerrero. However, La Cañada is most accessible from the Villa de Alvarez–Juluapan road, whereas La Parranda A is perhaps a hundred meters from the main paved road that connects the city of Colima with Comala. The turnoff is on the west side of the highway, and as one enters the property of La Parranda, the Capacha cemetery (La Parranda A) is immediately to the left, or south.

Essentially this is the same landscape as La Cañada, with great stretches of open, gently rolling country. An impressive view to the west, toward the Cerro Grande and Juluapan, has a break that provides a long-distance glimpse of San Palmar and of mountainous west Colima. La Parranda A is an exposed site in an open field, planted to maize every alternate year and allowed to lie fallow the rest of the time. Trees left along property lines and field boundaries include: *salate, higuera, guamuchil, parota, guásima, huicilicate,* and *higuerilla brava.* I do not have species identifications, but the first two are *Ficus;* the next three, *Leguminosae.* An extensive depression northwest of the cemetery has large trees beneath which coffee has been planted. There is no running water in the immediate vicinity, but three fine springs are situated about 1.5 km to the north.

The open fields are broken by low natural rises, many of which are said to contain looted shaft tombs. The Capacha-phase cemetery occupies a low, natural, gently curving rise (Fig. 2 *b*) about 100 m in diameter. It is dotted with a good many large boulders, including one pecked to produce what moneros consider a "map," supposedly recording the location of shaft tombs.

Several old excavations are evident, but the cemetery has not been exhausted simply because the soil is extremely hard and there is no demand for Capacha products. The monero who introduced me to the site had dug one large pit in which he seems to have found two burials, extended, accompanied by two bules.

I opened six pits (100 by 120 cm). Two (Tests 1 and 4) petered out almost at once. In one, subsoil was barely 20 cm below the surface; the other was dug to 100 cm, where the soft, yellowish, sterile soil, here called *peña,* was reached. The remaining four pits contained burials, although not exclusively of the Capacha phase.

Test 2

Approximately 100 cm below the surface, we cleared a welter of sizable stones, presumed to be associated with an interment. The stone heap ran north-south, but surprisingly, an underlying pit in subsoil had an east-west axis. Pit not undercut. Its floor was 155 cm below the surface and 140 cm in length. It contained a single Capacha interment (Burial 2; Fig. 3 *a*), whose only offering was a bule badly crushed by the stone. It was at the west end of the subsoil pit, with one tooth adjacent. When the pit was cleaned, another tooth and a tiny skull fragment came from the opposite end. No other bone; material too fragmentary to indicate orientation.

One sherd within the pit. Very small, it is of unslipped Red, unfortunately not classified. Paste quite unlike that of Capacha monochrome. A similar sherd found beneath the olla that accompanied Capacha-phase Burial 3 (see Test 5; Fig. 6, no. *6*). These two sherds suggest an early ware, not yet recognized, possibly contemporaneous with Capacha or perhaps even earlier.

Test 3

A non-Capacha interment (Burial 1, not illustrated) lay diagonally across the south end of a subsoil pit, with the legs extending into a cut made beneath the peña and presumably more recent than the original pit in which the upper half of the body reposed. Accompanying furniture aberrant, but probably of Colima phase; details not included here because of non-Capacha affiliation.

Bone collected as carbon sample 221 (UCLA 1651; Table 1, no. *6*) in the hope of a date for the Colima phase. One more recent than 500 years ago resulted, presumably the consequence of unexpected contamination.

Test 5

Here, we cut through top soil to a depth of 60 cm, below which was yellowish soil with many small stones. At the start of this yellow earth, we removed 6 or 7 fair-sized stones, horizontally placed. Below them were small fragments, apparently of two skulls (not assigned numbers).

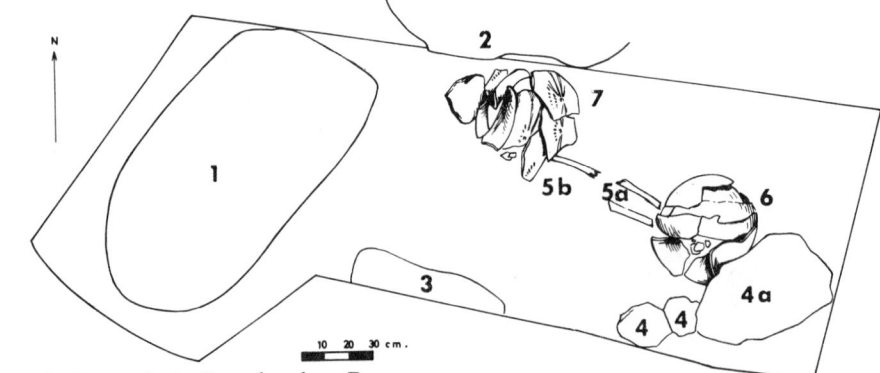

Fig. 6. Interment: La Parranda A, Capacha-phase Burial 3. *1, 2,* pits in subsoil; *3,* section of hard subsoil; *4,* stones; *5 a, b,* Burial 3; *6, 7,* pottery.

A pit in subsoil *(1),* starting at 135 cm below the surface and cut to 170 cm, was ill defined and empty, although in the fill 55 cm above it were small fragments, apparently of two crania (not shown and not numbered), but no other bone.

An old monero pit *(2)* abutted on our test; from it sherds of a painted variant of Capacha monochrome (Fig. 28 *d*) fell into our cut. *3,* an "island" or *barda* of hard subsoil; *4,* stones; *4 a,* a large stone that partially overlies the fragments of an olla of Capacha monochrome *(6).*

Burial 3 *(5 a, b)* consists of three fragmentary long bones; *5 a,* femora. Presumably the burial was extended, head to the northwest; it lay directly on subsoil but not within a prepared cavity cut into the latter.

Fragments of an undecorated olla of Capacha monochrome *(6;* Fig. 11 *i),* presumably placed on the legs of Burial 3 at the time of interment. Sherds of a decorated Capacha bule *(7;* Fig. 16 *a)* are directly on top of bone fragment *5 b.* (See pp. 24, 27, 43–44, 91.)

No indication of Capacha association for either stones or skull fragments. At 135 cm, sterile subsoil appeared; into it had been cut a poorly defined oval pit, 35 cm deep (Fig. 6, no. *1).* Pit quite empty; skull fragments mentioned above too near the surface to make association likely.

Original test was extended eastward, where we found Capacha-phase Burial 3 (Fig. 6, no. *5 a, b).* Burial rested directly on subsoil, but not within a prepared pit.

Sherd material from Test 5 very mixed. Capacha phase not represented in the opening cut, which included Ortices (1 sherd), Armería (4), near-Armería (4). The majority of classifiable sherds (55) were "late," with general resemblance to Chanal. But these were outnumbered by Red unclassified (50) and Plain unclassified (62).

From general digging in Test 5 there are 44 Capacha sherds. A figurine head (Fig. 31 *n)* reminiscent of San Gerónimo (coastal Guerrero) may belong in this or in later association. The few Comala (3), Colima (4), near-Armería (5), and "late" (vaguely Chanal) fragments (17) are far outnumbered by Red unclassified (61) and Plain unclassified (220).

Extension of the test over Burial 3 produced roughly the same mixture: "late" sherds (17), Red unclassified (81), Plain unclassified (183). In the immediate vicinity of Burial 3 there were Capacha sherds (3), Comala (1), Red unclassified (13), Plain unclassified (32). From immediately beneath the olla (Fig. 6, no. *6)* came a tiny sherd of unslipped Red, not classified; a similar fragment mentioned above from Test 2.

Test 6

Our largest and most productive test was at La Parranda A. Dark top soil to a depth of 60 cm, followed successively by soft, crumbly yellow earth; coarse gravel; and the sterile yellow soil (peña). The latter starts at about 170 cm below the surface.

This test contained two Capacha burials. From a pit in subsoil came Capacha Burial 4 (Fig. 7, no. *1),* accompanied by seven bules (Fig. 7, nos. *2–8).* Nearby were fragmentary remains of Capacha Burial 5 (Fig. 7, nos. *10, 10a),* apparently dislocated when Burial 4 was introduced.

Probably once associated with Capacha burials that were disturbed by subsequent interments were two bules (Fig. 7, nos. *12, 13).* These were not within subsoil pits and no bone was found nearby. A miniature jar, once tripod, with punctate ornament (Fig. 14 *e),* was loose in the fill near the pit that contained non-Capacha Burial 6 (Fig. 7).

The few sherds from Test 6 reflect use of the cemetery over a long period and include some Capacha monochrome, some Armería material, a few Red on brown or buff unclassified, and Shadow striped unclassified. As usual, Red and Plain unclassified are the most plentiful.

Within the pit containing Burial 4, sherds were almost exclusively Capacha monochrome. However, a small bit of Shadow striped unclassified fitted sherds of the same vessel, found when we cleaned the floor of the test. Apparently these belonged to non-Capacha Burial 7, part of which lay directly above the junction of bules 7 and 8 (Fig. 7) of Burial 4. A small sherd adjacent to said bules proved to be from a jar that accompanied Burial 7, indicating that this fragment, at least, had worked downward. There is no indication whatsoever that the Capacha repertoire included shadow striping.

Comments

Apart from providing a number of Capacha-phase vessels, the tests at La Parranda A had additional utility.

1. One of the few instances in Capacha association of a bowl with flaring rim comes from Test 5. The latter impinged on an old monero pit (Fig. 6, no. *2),* and loose soil from the latter spilled into our cut, bringing with it large fragments of the bowl. The latter also is of interest in being one of the few cases of unzoned red on brown decoration from the Capacha phase (Fig. 28 *d).*

2. From general digging in this same test came a figurine head (Fig. 31 *n)* reminiscent of the Guerrero coast (cf. Vaillant and Vaillant 1934: Fig. 17, especially *k, l).*

4. TERRENO DE JESUS GUTIERREZ

Concerning the Capacha cemetery on the Terreno de Jesús Gutiérrez (Fig. 1, no. *4)* there is little information. The paved road from Colima to Comala passes through Villa de Alvarez, where it turns sharply northward and

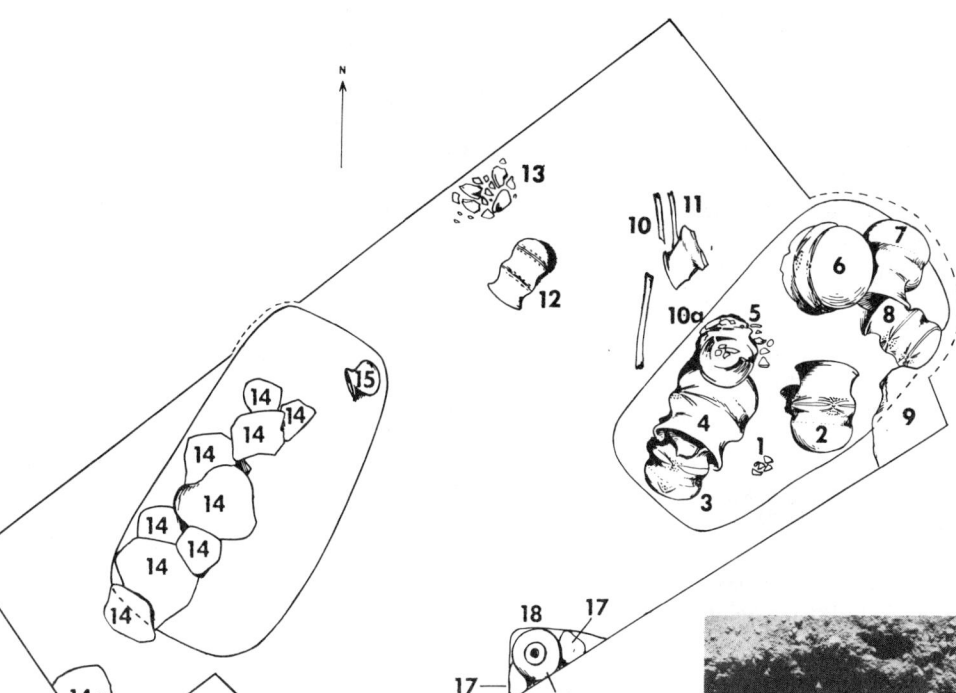

Fig. 7. Interments: La Parranda A, Capacha-phase Burials 4 and 5, and non-Capacha interments 6 and 7. *1*, Burial 4; *2–8, 11–13, 15, 16,* pottery; *9, 14, 17,* stones; *10, 10 a,* Burial 5; *18,* subsoil pit. (See pp. 24, 43–44, 91–93.)

Burial 4. Bone is confined to small bits of a cranium *(1)* on the floor (220 cm below the surface) of a shallow, well defined pit cut in subsoil. Accompanied by seven bules of Capacha monochrome; *7,* undecorated, the others with typical broad-line incision and punctation. Except for *3* (Appendix IV–A–3, 9609; Figs. 15 *d;* 17 *d),* none is illustrated: *2* (Appendix IV–A–3, 9613), *4* (IV–A–3, 9610), *5* (IV–A–3, 9611), *6* (IV–A–3, 9616), *7* (IV–A–3, 9614), *8* (IV–A–3, 9615). *9,* a large stone in the corner of the test, abutting on bule *8.*

Burial 5 consists of three fragments of long bones *(10, 10 a,* the latter a femur shaft that appears unusually long, measuring in situ, without either articulation, 44–45 cm). The only accompanying furniture is the rim of a large Capacha bule *(11,* not otherwise illustrated), 170 cm below the surface. Apparently Burial 5 was destroyed when the pit for Burial 4 was cut. Note that Burial 4 is contained within a prepared subsoil pit, while Burial 5 has no such excavation. The photograph shows a partial view of Burials 4 and 5 before bules *7* and *8* had been uncovered.

Outside the burial pit, with no bone associated, is a small bule of Capacha monochrome *(12),* with atypical incision (9617; Fig. 20 *a);* it rests on subsoil, at a depth of 170 cm. Fragments of another bule, with normal decoration *(13;* not otherwise illustrated; Appendix IV–B–3, 9617 a) are nearby, but at higher level (135 cm below the surface).

There are several non-Capacha interments (including Burials 6 and 7) in this test. Burial 6 is represented by fragments of both legs on the floor of a pit at 220 cm below the surface and cut into subsoil which here starts at 170 cm. The bone is hidden in the figure by a series of large stones *(14),* above it but at higher level (90 cm below the surface). Stone *14 a* probably was not related to this interment. The only offering associated with Burial 6 is a small jar of unslipped Red unclassified *(15;* 9619, not listed in Appendix IV because it evidently is post-Capacha).

Burial 7, similarly unplaced as to phase, but also post-Capacha, is not shown. It is represented by the fragments of a small, upright olla of Red unclassified (9618, not in Appendix IV) and a few bits of skull. These were found 120 cm below the surface, immediately above the point where bules *7* and *8* impinge. There, the northeast wall of the test was undercut, to follow the line of the subsoil pit and to free bules *6, 7,* and *8.* The overlying burial was not cleared, although the accompanying vessel was removed.

One more non-Capacha burial, possibly two, were found on the fringes of this test. In the southeast wall of the cut was an upright jar of near-Armería cream ware (*16,* 9622, not in Appendix IV), its mouth 150 cm below the surface. Flanked by two medium-sized stones *(17),* the vessel sat immediately atop a small effigy tapadera (not visible, 9623, not in Appendix IV), at 190 cm below the surface. Doubtless the two pieces are related, although most tapaderas occur in Colima rather than Armería association (pp. 8, 9). Deeper, we came upon the corner of a subsoil pit *(18),* at 210 cm below the surface, with the floor of the cavity at 230 cm. Several small stones were removed and a few splinters of bone appeared beneath them. This may or may not represent another interment. No burial number was assigned and no further clearing attempted.

then proceeds in a straight line to Comala. As one leaves Villa de Alvarez, beyond the curve, a great expanse of level fields lies west of the road; to the east of the road a few houses continue on the outskirts of town. Here, perhaps 300 m north of the curve and 100 m west of the road, I noticed, quite by chance, a surface scattering of Capacha sherds. A monero-guide was taking us to a site farther west, and we cut across the fields at this point. Now that the surface sherds have been collected, it may be difficult to locate the cemetery, especially since industrial construction has begun in this area. In any case, the cemetery is a few meters south of an incipient arroyo, the latter perhaps 1.5 m deep and apparently connected with the highway drainage. Otherwise, there seems to be no stream, ancient or modern, in the immediate vicinity.

It turned out that the monero-guide himself had dug exploratory pits here some time before. He reported having found three burials, each with a typical bule to one side of the head. He thought the heads were to the west, and each skeleton was "in its box," referring to a pit in subsoil. Upon another occasion, two of his acquaintances dug here, likewise finding a few bules. Inasmuch as the latter had no value, the moneros are said to have smashed the vessels against the trunk of a large huisache shrub.

Fragments we collected on the surface come from at least three bules, one semirestorable. Our one carbon date for the Capacha phase (Table 1, no. 14) is based on surface sherds collected at this cemetery. Other surface fragments —not burial furniture—include Ortices and Comala wares, plus Amoles wiped;[3] in other words, they are post-Capacha.

5. PARCELA DE LUIS SALAZAR

Description and Tests

The Parcela de Luis Salazar (Fig. 1, no. 5) is about 4 km northeast of the city of Colima, adjacent to the dirt road to El Chanal; the latter is both an ejido and a federal archaeological zone. The Capacha cemetery is about half a kilometer downstream from the gate that gives access to the archaeological zone and lies outside it. Even so, it is within the limits of the El Chanal ejido, not far north of its boundary with the property called Santa Bárbara.

In recent times, the plot in question has been planted to papaya trees; they have not prospered, perhaps for lack of water, perhaps because surface soil is shallow. Although the Río de Colima passes the site on the west, it is entrenched several meters, so its waters are not available for irrigation. Moreover, the river supplies the city of Colima with water, and by El Chanal the course is all but dry.

I visited this small papaya plantation in 1970 and collected, in an area about 50 m square, surface sherds that resemble those of El Chanal—that is, material presumably late in the Colima sequence. No Capacha fragments were included. Nevertheless, the informant-guide reported that some seven years earlier he and three companions had dug here, finding about six bules—some red, some black, some with double cincture. He indicated a spot 15 by 15 m as the source of the Capacha material; later, he was inclined to enlarge the area to 15 by 23 m. One of his former collaborators indicated an extension somewhat to the south.

In any case, the owner refused to permit excavation because of his papaya planting. By the following year, however, the trees were in such deplorable condition that their value obviously was negligible and, reluctantly, he agreed we might test.

Tests 1 To 4

Small pits (100 by 125 cm) were opened on the presumably undisturbed fringes of the area said to have produced burials accompanied by bules. Subsoil at uneven depth, varying from 50 to 100 cm.

Sherd material mixed: some Comala, some Chanal. Red and Plain unclassified outnumbered identifiable fragments. Such an unaccommodated block of Red and Plain is uncomfortably common, especially in the northern part of central Colima, and very much needs study. Tests unproductive from viewpoint of Capacha phase.

Tests 5 And 6

Moneros agreed that very few bules had been found entire and that all fragments had been abandoned on the spot. In the hope of finding a Capacha interment overlooked by them or at least, their Capacha discards, we moved into the plot that informants said already had been dug.

Here we found remains of two, possibly more, fragmentary burials, one with Capacha furniture. These were within 60 to 80 cm of the surface, resting directly on tepetate but not within a cavity cut into it.

Burial 1 consisted of a skull, head to east, face to west; mandible out of alignment, some centimeters to northwest. Body apparently supine, probably with fragments of restorable cauldron of Capacha monochrone (Fig. 12 c) on chest. Said vessel contained a miniature olla, extremely thick, but presumably Capacha (Fig. 14 d). Offerings equidistant from skulls of Burials 1 and 2 and could have belonged to latter.

Burial 2 was a lone cranium, just east of Burial 1 and, like it, with head to east, face to west. One bone fragment, a humerus, found beneath the Capacha cauldron noted above, may represent remains of a third individual; axis does not suggest identification with either Burial 1 or 2.

One meter south of Burials 1 and 2, we found two large fragments of Capacha bules, adjacent to a heap of stones. No trace of bone. Unfortunately, at this point, the owner of the parcel sent a messenger to inform us he had given permission for only two days of work. Although this had not been my understanding, rather than dispute the matter, we simply closed the several pits.

Comments

Testing at this parcel was not productive, but at least it confirmed the existence of burials of Capacha affiliation on the southern outskirts of the El Chanal site.

6. LA CAPACHA

Description and Tests

La Capacha (Fig. 1, no. 6) is an ex-hacienda which, in recent years, has become in large measure an ejido, with land parcels in the hands not only of local residents but also of ejidatarios of Comala who could not be accommodated closer to the latter town. Accessible by dirt road from the city of Colima, the small cluster of houses at La Capacha is about 7 km northeast of the state capital and possibly 2 or 3 km east of the archaeological zone of El Chanal.

This is attractive country, in part open, in part with low natural hills. It is dotted generously with venerable *parota* and *Ficus* trees. Water from a local spring is supplemented by that brought by a small canal. Maize is planted during the rainy season, in alternate years; there also is some irrigated maize. Dairying is important. As is the case with neighboring El Chanal, La Capacha must have invited human habitation from early times, and archaeological evidence to this effect is abundant.

The Capacha cemetery proved to be a plot not more than 25 or 30 m square, situated in the great expanse of land known as the Potrero del Casco. Immediately to its west is a small wooded area known as the Bosque del Huicilicate and, until the 1950s, this stand of trees extended eastward and covered the Capacha burial ground. At that time, the latter area was cleared, to be planted to maize. When not under cultivation, it is covered with the thorny shrubs called huisache and *garabato negro* (Fig. 2 *d*). The cemetery area slopes gently eastward, where fields at somewhat lower level are given over to sugarcane.

In 1969 the survey was extended to La Capacha; surface sherds were collected from various localities. My monero-guide was unaware of any pottery answering the description of a bule, but one day we called on another local resident who displayed, unenthusiastically, some material he had excavated a few days before. Included were a small unmistakable bule and several artifacts plausibly said to have been found in association with it (Appendix IV-C-6, 8679–8687). Amiably, the new informant took us to a nearby field where he had a second Capacha burial partially exposed. I was present when the grave furniture (Appendix IV-A-6, 8676–8678, 8706–8708) was removed, and, while this hardly represents controlled excavation, association is secure. This material was purchased on the spot, likewise that which had been extracted a few days earlier. The monero said he had found a total of four burials, presumably all Capacha.

Because of commitments elsewhere, I was not able to test this Capacha cemetery until the start of the following season. Results were disheartening. During the several intervening months. the first monero-guide took over the plot and, according to his account, found two additional interments. Fortunately, he was unable to find an interested dealer, and I purchased the material: one alleged grave lot (Appendix IV-C-6, 8712–8714) and a second lot said to have come from the immediate vicinity, although not associated with an identifiable burial (Appendix IV-E-6, 8715–8720 a–k).

My series of small pits at this cemetery demonstrated clearly (1) that it had been effectively exhausted and (2) that the sterile soil known as tepetate is almost literally at the surface. Some of the tests were not even numbered, for top soil ended at a depth of 15 cm. In two natural crevices in tepetate, between 90 and 100 cm deep, post-Capacha sherds, possibly from the adjacent site within the Bosque de Huicilicate, had accumulated.

We found no burials but were able to salvage a few isolated Capacha-phase specimens that had escaped the moneros. These artifacts are of considerable interest:

1. Small Capacha monochrome bule (Figs. 9 *j*; 16 *d*).
2. Small Capacha monochrome olla (Figs. 10 *f*; 22 *a*) with incision and punctation. Not restorable; so near the surface that part of the rim probably had been carried away in plowing.
3. Stone "plate," shallow (Fig. 42 *b*).
4. "Spoon" or "ladle" (Figs. 42 *a*; 43 *c, cc*).

Capacha association for these pieces is secure and is particularly welcome for the stone artifacts, because it lends credence to the claim of such affiliation for several stone manufactures purchased previously from the two local moneros.

The cemetery at Capacha and the immediately adjacent field are dotted with several large basalt stones, some of which rest directly on tepetate. Two of these stones have winsome human faces crudely pecked. The easterly stone, on the edge of the burial ground, has a face on its western surface; another stone, 5 or 6 m distant, is larger and has three human faces on its southern and three more on its northern side. We cleaned the base of both stones and found tepetate almost at the surface. Against the southern face of the larger stone, but not beneath it, was just one sherd. It is the fragment of an annular base—a form infrequent in Colima—and much too recent to belong to the Capacha phase, but it might relate to the Bosque del Huicilicate material. It cannot be said definitively that the stone faces are Capacha or that they are more recent; as a guess, the latter seems more likely.

Comments

The Capacha phase derives its name from La Capacha, whence came the first material to permit clear, albeit provisional, recognition of a discrete ceramic assemblage—one with a variety of pottery forms and with assorted artifacts, particularly well-worked, ground-stone products, in association.

From elsewhere in the vicinity come tantalizing suggestions of additional Capacha material. Some time ago, (1) two "long black" vessels were found in excavations at a small, mixed site I am calling El Zopilote *bis*. Also what from description seem to have been bules were reported from (2) a burial "opposite" (presumably across the road from) the Huerta de Capacha, from (3) a spot near Volantines, and (4) from one at Guerrero, near the Potrero de Bazán. Finally, (5) a vessel, possibly a bule, was found in a "small tomb" in the vicinity of Potrero Duro. All these reports were investigated but without concrete results. The whole area about La Capacha has been sacked so mercilessly that it is unlikely any considerable concentration of Capacha interments has escaped discovery.

7. EL BARRIGÓN

Description and Tests

Buenavista (once called El Tecolote) is a sizable ejido that grows rice and sugarcane (Fig. 1, no. *7*). It is low on the flank of the Volcán de Fuego in open, frost-free country, 11 or 12 km south of Cuauhtemoc (the former San Gerónimo) and immediately west of a series of small streams that unite to form the Río Salado. A surface scattering of badly eroded sherds attests ancient occupation in the vicinity of these headwaters, but the only site that seems sizable and apparently important is one called El Barrigón (Fig. 1, no. 7). It lies in cane fields, about 3 km northeast of the ejido settlement of Buenavista and somewhat west of the modest Arroyo de la Huerta, which appears on some maps as the Barranca de San Gerónimo.

El Barrigón is large by local standards, with a north-south extension of 800 or 900 m and an east-west spread of perhaps half that figure. Over this expanse, low mounds are continuous, many almost imperceptible and none more than 3 m in height. The entire surface is carpeted with water-worn stones (Fig. 2 c), presumably hauled from the nearby arroyo bed; they seem once to have been concentrated on the mounds but have been thoroughly scattered as a consequence of plowing. On the north, low east-west rises alternate with shallow depressions that might be borrow pits. Mound arrangement is blurred, and there seems to be only one cluster of rises that might possibly have formed a plaza. A small test suggests that the mounds are formed of earth and river boulders. At the south end of the mound area there is a marked "trough," which workmen call a *canoa* (canoe) or a *calle* (street); it is 80 or 90 m long and about 20 m wide, and has a tallish mound at the western termination. The proportions suggest a ball court. This mound area, just described, is not of immediate concern, because it is presumably post-Capacha.

Prehistoric Canal

El Barrigón deserves special mention as the only locality in Colima where a well-defined pre-Conquest canal has been observed. This unusual feature was reported by a monero, who described a "trench" one *vara* (yard) wide, covered by large stones tilted against one another to form a gablelike covering. Because of the latter, the channel was not filled solidly with earth, and the man said he had inserted a crowbar, which indicated a considerable linear extension, roughly northeast-southwest.

At the time, I was interested chiefly in Capacha-phase burials, but the report of a prehistoric canal necessitated confirmation. Some 10 m southwest of the monero's pit and in line with it, I opened a small test and there found the canal, 85 cm wide, cut tidily into the hard subsoil tepetate, which here is 1.5 m below the surface. Depth was not determined, for the cavity was filled with large river boulders, which the workmen extracted inexpertly and with difficulty. Neither gablelike cover nor stones suitable for such arrangement was evident.

The alignment suggests that the central area of the mound assemblage once was connected by canal to an upstream point on the Arroyo de la Huerta. The latter is now deeply entrenched, but a millennium ago it may have been a shallow stream, perhaps fed by large springs. To trace the canal to its source and to try to date it would be a comparatively prolonged job. Few sherds came from the test over the canal; the one identifiable one is Amoles wiped.[3]

Other Post-Capacha Features

The second unusual aspect of the El Barrigón site is the long, narrow depression mentioned above. Here I opened a test (75 by 130 cm) within the depression and close to the rise along its southeast side. Some large stones presumably had formed part of the bounding ridge, but there was no indication of a formed wall, either of masonry or of other construction. From this entire test, not more than five or six Plain unclassified sherds resulted, these exclusively from the uppermost level. The top soil (to 30 cm) was yellowish and hard; below was a dark band (30–70 cm); from 70 to 200 cm again the soil was yellow; and at 200 cm the underlying tepetate began. Resting on the latter were several fairly large flat stones, but there was no suggestion of a prepared floor.

Capacha Cemetery

On the southern fringes of the mound area, on the almost level plain, there was a Capacha-phase cemetery, which presented the first opportunity of finding Capacha material in situ. In the mid-1960s several shaft tombs, certainly not Capacha, were found here, and, according to informants, the day after the discovery some 60 men were digging enthusiastically in the vicinity. Participants guessed that this labor force destroyed between 60 and 80 Capacha burials, but inasmuch as the accompanying offerings were not marketable, little attention was paid to them, much less to skeletal material. Children smashed the pottery with sheer pleasure in the destruction; some bules were taken to the village, to be broken as piñatas.

The entire field had thus been churned, and with one exception, my tests were in disturbed soil, with small, unarticulated fragments of human bone common. Only two pits require specific mention.

Test 5

Pit almost square (140 by 170 cm); aligned northeast-southwest. At 90 cm below surface, a break in underlying tepetate crossed pit diagonally, from north to south; apparently the line of an ancient excavation, which we could trace only on its eastern side. On floor of depression, 120 cm below the surface, there were fragments of several long bones, not aligned; south of them was a mandible, underside upward, together with a skull fragment. Some suggestion of an extended burial, head to south, possibly prone. Break in tepetate may be all that remains of a former pit cut in subsoil to receive a Capacha interment (see Chapter 3: *Interments*).

Forty cm south of mandible, in extreme southern corner of pit, a fragmentary cranium, unaccompanied, in soft soil; latter apparently the false start of the shaft for a tomb, left incomplete.

No burial furniture in either of the above cases, hence phase allocation unknown.

An extension 1 m square northeast of Test 5 included a small area overlooked by moneros. Here we found a Capacha bule (Fig. 19 b) upright, its rim 30 cm below the surface, its body in direct contact with the face of a badly smashed skull (Figs. 46, 47; Appendix V, no. 8632); small stones at base of vessel held it upright. Evidently the remains of a burial, head to south, face looking northwest.

No mandible, but bule on the very edge of a monero pit that had destroyed the rest of the skeleton. One small obsidian chip but no sherds in immediate vicinity; bule identifies interment as unequivocably Capacha.

Test 6

Situated 1.5 m south of original cut of Test 5. One cranium fragment with part of mandible on top, plus scattered bits of bone. Fragment of stone ax with three-quarter groove (Appendix IV-B-7, 8890) and, from impinging monero pit, the lower part of a heavy Capacha monochrome vessel, evidently with subconic neck (Fig. 10 *d*; cf. Eisleb 1971: no. 192). Not a long, narrow-necked bottle, but an approximation to bottle form.

Comments

El Barrigón as a whole—mound area and Capacha cemetery—presents a wide range of ceramic material, albeit sparsely. The few sherds available from the mounds are primarily Red and Plain unclassified, which must be studied carefully and allocated to phase before the mounds can be seen in perspective. However, these unclassified sherds were accompanied by a few fragments reminiscent of Comala red, and the rim form of some of the Red unclassified rather suggests the ill-defined Manchón pottery (Kelly 1978). There is a minor but persistent occurrence of Amoles wiped ware, which elsewhere may be found with Ortices, or with Comala, or unaccompanied.[3] No Ortices sherds were included in the small lot from the mounds, although some cropped up in the tests in the adjacent Capacha cemetery. At the moment, the best guess is that the mound area is post-Ortices, perhaps contemporaneous with Comala.

The tests in the flat, almost stoneless cemetery area were far from uniform. Tests 1 through 4 related ceramically to the mound material, not to Capacha. Test 5 produced seven Capacha sherds, outnumbered by 38 Ortices sherds. From Test 6 came a few Capacha fragments, together with a scattering of Ortices, Comala, and Amoles wares. In contrast, Test 7 produced almost exclusively Capacha pottery (44 sherds). Inasmuch as the yield was scant and the chances of stratigraphy in the thoroughly churned cemetery were nil, formal sherd counts are not enlightening. Although the Capacha area was completely rifled, the vicinity of the mounds is still relatively undisturbed and should repay systematic excavation; traces of Capacha and Ortices well may underly the more recent occupation.

In summary, El Barrigón provided several bits of useful information concerning the Capacha phase. The small surface collection made in 1968 included two painted sherds, unique in the Colima collections. They are purple-black in combination with rose red, the colors separated by a wide groove (Fig. 31 *g*). Informants claim to remember several entire vessels of red and purple-black from the rifled Capacha burials; I know only the two sherds just mentioned.

One unglamourous, semirestorable vessel, with subconic neck, approximates a bottle in form (see Appendix II: *Form: Cántaros*). Owing to the mixed sherd content of Test 6, it is impossible to attribute its ax fragment securely to the Capacha phase, but such allocation seems likely (see Appendix III: *Axes*).

8. TERRENO DE FIDEL VALLADARES

Description and Tests

Further indication of Capacha material comes from Terreno de Fidel Valladares (Fig. 1, no. *8*), in the Río Salado drainage, some 20 km downstream from El Barrigón, in the vicinity of Acatitán. The latter is an ejido settlement on the open, mesalike area just west of the main Salado stream and east of the head of the great barranca which, a few kilometers downstream, passes Los Ortices on the east.

During the 1960s water for irrigation was brought to Acatitán, and in 1970 a small ditch to convey it to certain fields was cut across the property of Don Fidel Valladares (Fig. 2 *e*). From this excavation, the latter's son, Don Enrique Valladares, salvaged the fragments of a handsome stirrup pot (Figs. 24 *b*; 25 *c*), reminiscent of a specimen in the Museum of the American Indian, Heye Foundation, and presumed to be a Capacha product (Kelly 1972: Fig. 32 *d*). It was reported that previous excavations in the immediate vicinity had turned up a number of burials accompanied by large vessels called *jarrones*, the local name for bules. A horizontally compound vessel, consisting of three united plates, was mentioned; also an effigy pot, from description apparently similar to that of Figures 13 *f* and 26 *f*; and a "small metate."

In the spring of 1971, I visited the Valladares property, located at the southern limits of the ejido and quite fully cultivated. About 100 m east of the dirt road that runs from Acatitán to Las Borregas is a masonry building in disrepair, one of the several relics of hacienda days. Between this structure and the road, slightly to the south, is a very low natural rise, almost bare of vegetation when I saw it (Fig. 2 *e*), but planted to maize during the rainy season. This was the scene of the Capacha burials.

Over the years, the cemetery has suffered badly. In addition to the narrow irrigation ditch, there have been excavations made to obtain sand for construction purposes, starting in hacienda times and continuing to the present. The original dimensions of the cemetery are by no means clear but may have been approximately 40 by 40 m.

The irrigation ditch runs roughly east-west. Using it as an axis, I cut a total of 18 pits parallel to it—those to the south, 140 cm removed; those to the north, 125 cm. Tests were 75 by 200 cm and were spaced every two meters. The top soil is hard; at about 40 cm sand begins and continues to 60 or 70 cm; then comes sterile gravel.

South of the irrigation channel we found two interments (Burials 1 and 2) and to the north, one (Burial 3).

Burial 1

Small child; remains limited to a few bone fragments, some tiny bits of skull, and four deciduous teeth. Apparently the head had been to the south. Interment in gravel, 95 cm below the surface; not in any discernible pit. The only offering was a small green-stone bead (Fig. 35 *f*); burial not assignable to phase; as a guess, Capacha.

Group Burial 2

A multiple burial, apparently secondary, contained in an oval pit cut in gravel. Pit began at 70 cm below the surface and was 15 cm deep. Its length of 125 cm not fully utilized; a confused welter of bones heaped along east side and at south end of pit. As far as I could tell, two adults and possibly three children represented. No furniture; half a dozen sherds of Capacha monochrome in association.

Burial 3

Partial interment, legs alone; upper part of skeleton within area destroyed by irrigation ditch. Body placed within a shallow oval pit, 90 cm below the surface, cut slightly into gravel. Head had been to the south. On top of the feet, a pile of stones, reaching almost to the surface; included were one mano, a fragment of another, and part of a deep, very narrow metate, with well worked exterior (Fig. 37 *c*).

We found no pottery, but the stirrup vessel mentioned above was from this area and probably accompanied Burial 3; not found when the ditch was cut but exposed later, when the channel was in use and the fragments were uncovered by running water. Bank of channel eroded in vicinity of Burial 3, which may represent spot from which vessel was washed out; at time specimen was exposed, no bone was observed.

Comments

All told, this cemetery was a great disappointment. Despite 18 tests, we found no interment even semicomplete and none with identifiable furniture. Although scant, surface material and sherds from the tests are overwhelmingly Capacha and evidently indicate a Capacha cemetery, now totally destroyed.

9. LAS BORREGAS

Comments

Acatitán is bounded on the south by the ejido of Las Borregas (Fig. 1, no. *9*). Just over the dividing line, in Las Borregas territory, is a great expanse of open maize field, the property of Don José García. I heard, in 1967, that a series of tombs had been excavated here, at least one of which had produced black pottery; an informant described a black jar with a face in relief on the neck. Two large vessels, evidently bules, were sold; the remaining pottery was left within the tomb and the latter filled. The informant vaguely remembered some red bowls, to which he paid little attention; he was sure there were no figurines and no stone artifacts. His account was confirmed by his brother, who had collaborated in the excavation. On the surface, I found a couple of sherds that attested the presence of at least one Capacha bule. At that time, the Capacha phase still was not recognized, nor its importance appreciated.

In 1971 I returned to Las Borregas, hoping it might be possible to find and reopen the tomb. The informant declared that the tomb was entirely normal for the area— with shaft, a drop to the chamber floor, and a sealing stone. However, he felt there was little chance of locating the filled tomb, for the maize field was large, many other tombs had been excavated close by, and continuous cultivation had blurred surface remains. He added that the real reason the tomb was abandoned was that it had been dug

in hard sand, and he and his brother feared it might collapse on them. About all that can be said is that this is a very convincing report of a shaft tomb with Capacha vessels, the identification backed by a couple of surface sherds.

Still another account of a shaft tomb with Capacha material was recorded in Las Borregas. I was not able to visit the locality, which is called El Ciruelo; it is said to be near the Río Salado, on its west side, not far from the point where the trail going east crosses the river.

My knowledge of El Ciruelo dates from 1966, when, in the ejido settlement of Las Borregas, I saw two large bules, undecorated, in use as flower pots. One had a single cincture; the other, two. A member of the household said that a black mask had been found with these specimens, as well as a vessel with concave sides, incision, and "three or four feet." Masks, black or red, are associated exclusively, as far as I know, with the Comala phase, and the footed, concave-sided pot might also have been Comala. Unfortunately, none of the pieces except the bules were available for inspection, and the owner was not at hand. The following season, I talked briefly with him. He said that the bules had indeed come from a tomb, together with black ollas and black masks; there was no red pottery. In 1971 I happened to chat with a local resident who had been present when the El Ciruelo tomb was opened. He remembered one bule and one mask and added that the tomb had a sealing stone, but he was uncertain whether there had been a drop from the floor of the shaft to that of the chamber.

It is said that El Ciruelo has been excavated quite thoroughly, although the earth is so hard that not many are interested in working there. Unfortunately, at the time of my visits, no guide was available, and Las Borregas, which seems to have two convincing reports of Capacha products from a shaft tomb, awaits further investigation.

10. QUINTERO

Description and Tests

A short distance downstream from Las Borregas, the Barranca de Ortices enters the Salado, through a gorge so rugged and deeply entrenched that the area is known as the Espinazo del Diablo. Apparently this chasm, which resembles a miniature Grand Canyon, has been selected as a dam site. Shortly below the gorge, the Río Salado emerges in a narrow and irregularly shaped valley. The river shifts course frequently from east to west and, since the 1959 hurricane, its channel has been so wide that it occupies a good deal of the valley floor, which once was arable land. At present, the stream runs against the base of the cerros that bound the valley on the west.

The small amount of flat agricultural land in the valley today is in pockets east of the river, and planting is largely on steep cerro slopes. There is one maize harvest a year, and although the present population is quite small, in recent years people have tried to move elsewhere, in search of better agricultural potential. The two small modern settlements in this restricted stretch of the Salado valley are the ejido of Las Trancas, just below the point where the river emerges from its gorge, and Caután a few kilometers downstream. In passing, it may be noted that Acabtlan or Acautlan was the name of a sixteenth-century native pueblo, located somewhere in this general area (Sauer 1948: 47, Map 3, Acabtlan). Both Las Trancas and Caután are on the east bank of the Salado, almost on the edge of its present course.

Likewise east of the river, halfway between Las Trancas and Caután, is a stretch known as Quintero (Fig. 1, no. *10*). This is not a settlement, but there is a corral for stock, and somewhat upstream one inhabited hut is situated on the valley floor, adjacent to and barely 2.5 m above the flood plain. The latter tends to be swampy and is little exploited, although occasionally a risky banana planting occupies part of it. Some 30 m downstream from the hut just mentioned is a small Capacha cemetery, unusual in being on comparatively low land. The surface soil is hard-packed, sandy silt, which merges gradually into sand and, at about 120 cm below the surface, into gray river sand and gravel.

The area of the cemetery is under cultivation sporadically, but at the time of our visit was fallow. Secondary vegetation consists of thorny huisache, *coliguana, zapotillo, cacanahual,* and dry grass. An enormous parota tree stands between the hut and the burial ground. Just east of the latter, and separated from it by a dry-stone wall, is the old north-south, dry-weather road from Ixtlahuacán to Las Trancas. When the road became eroded, traffic began to run parallel, a few meters to the east, forming the present road. The old one now is an incipient arroyo (Fig. 2 *f*) in whose banks human bones are occasionally exposed during the rains. The cemetery proper may not be more than 15 or 20 cm square; we tested there, as best we could, among monero pits, and also in the extension eastward, beneath the old road.

A local monero reported that in the center of the cemetery burials were concentrated and accompanied by offerings; interments on the fringes generally were without furniture. This information may be correct. Our pits in the central area produced a good many isolated bone fragments, suggesting sequential interment, with the later ones destroying previous ones. Here also we found several entire Capacha vessels, probably relics from burials destroyed when the plot was reused.

The monero-informant reported having dug a large pit in the central part of the cemetery; he said it contained four extended burials: two with heads to the west; one, to the east; one, to the north. A considerable amount of grave furniture was reported, from description exclusively Capacha. By the end of the 1970 field season, we had cleared Burials 1–6.

In the spring of 1973, I returned briefly to Colima, in the hope of obtaining additional data concerning the Capacha phase. Quintero was revisited, on the chance that one or two more pits in the arroyo bank east of the old road might be productive. It turned out that a large-scale program of road improvement was under way. The old road had been destroyed completely and the adjacent new one was about to be surfaced with cobbles. Informants said that some bone fragments had been found during the

road work, but apparently no recognizable burials and no artifacts. I made one test just east of the new road, but the soil was free of archaeological remains.

Burials 1–6

In the heart of the cemetery, we found one interment (Burial 1), 80 cm below the surface; extended; head to north; prone. No offerings, hence phase allocation uncertain; bone saved (carbon sample 238), in the hope of a radiocarbon date.

East of the stone wall and along the west edge of the old road, now an incipient arroyo, were fragments of a cranium (not numbered); head to west, supine; entire body had been carried away by the arroyo.

Burials 2 through 5 (the last mentioned, Appendix V, no. 9752) in the bank that bounds the same shallow arroyo on the east. Burial 4 (Fig. 8, no. *2*) of special interest because it produced a stirrup pot with two tubes (Fig. 25 *a, aa*). Burial 6 (not illustrated), 1 m southeast of Burial 5, at a depth of 90 cm; extended, head to north, face down, arms at sides. No offerings; bone saved as carbon sample 241.

In southwest corner of test containing Burial 6, and absolutely flush with the surface, two large sherds of non-Capacha pottery. One was the upper part of a sizable cántaro, rose-red, with widely flaring rim and with finger grooving on the upper body. It rested within the fragments of a rose-red bowl, with black interior and with slightly incurving sides. As yet, these two specimens not placed as to phase. Presumably, they are all that remained of a post-Capacha interment. No bone accompanied the sherds, but the entire skeleton would have been within range of the arroyo.

Burials 7–10

Adjacent to and beneath the level of an "oven" (see below), although not actually underlying it, we found Burial 7, resting on gravel, the top of the cranium 120 cm below the surface. Head to north; face may have been to south or southeast. No bone other than the skull (Appendix V, no. 10300). An undecorated bule (not illustrated), deliberately "killed," was where the mandible should have been.

Concurrently, 10 m to the north, we cut into a small area that looked intact, although surrounded by monero pits. At 50 cm below the surface, a bule rim appeared; vessel (Fig. 17 *i*) rested on silt, immediately above the start of underlying sand. On the assumption that this was an interment, it was identified as Burial 8 (Fig. 3 *b*); later, it turned out that no bone accompanied the bule.

Immediately above the latter and almost in direct contact with it was a fair-sized comal sherd, complete with perforation to form rim handle (Fig. 31 *m*). Other comal sherds (see Appendix II: *Griddles*) came from adjacent fill, as did one fragment of a prismatic obsidian blade and one of a slablike figurine of a style unknown to me. This cluster of items is presumably post-Capacha.

Burial 9—again, assigned a number in anticipation of a skeleton that did not materialize—from the same pit as Burial 7. Represented by a small undecorated bule with

Fig. 8. Interments: Quintero, Capacha-phase Burials 3, 4, and 5. *1,* Burial 3; *2,* Burial 4; *3, 5,* pottery; *4,* Burial 5; *6,* milling stone. (See pp. 24, 50–51, 92.)

A series of fragmentary interments in a depression (about 70 cm deep), which formerly was the local road, now an incipient arroyo.

Burial 2, not shown, has been removed. It was in poor condition, with fragments of the cranium overlying a stirrup-mouth vessel *(3).* The head was to the north, face upward, body extended; teeth quite worn. No offerings, hence this interment cannot be identified as to phase. It was barely 50 cm below the present surface.

Burial 3 *(1)* lies in part beneath Burial 2. It consists of a few skull fragments, a mandible, and a couple of long bones. Head was to the north, face to the west; at 80 cm beneath the present surface. Again, no furniture, but possibly of the Capacha phase.

Burial 4 *(2)* is immediately west of Burial 3 and 20 cm deeper. Fragmentary, but evidently it was extended, with the head to the north. The head has fallen on the chest, so that the top of the skull is upward; no trace of face or mandible.

It is uncertain if Burial 3 or 4 is the earlier. A Capacha monochrome vessel, with stirrup mouth *(3;* Figs. 13 *d;* 25 *a, aa)* presumably accompanies Burial 4; it is close to the skull and at the same level. Hidden and almost beneath the stirrup pot is a small, thin plaque of unfired marine clay (p. 77; not illustrated; Appendix IV-A-10, 9746). Six obsidian chips (not shown; Appendix IV-A-10, 9744 a–f) came from beneath the stirrup vessel, and an obsidian scraper (not shown; Fig. 36 *e)* lay 15 cm to the east, at the same level. It may or may not have been part of the furniture of Burial 4, but Capacha association seems secure.

Child Burial 5 *(4* on sketch; Appendix V, no. 9752) is south of Burial 3 and at 20 cm lower level; extended, with head to the north, face to the west. Evidently it predates Burial 3, for one of the bones of the latter lies above the skull of Burial 5. Furniture of the latter includes an olla of Capacha monochrome *(5;* Figs. 11 *j;* 22 *b;* 23 *a)* at the chest and, to the rear, in the lumbar area, a well worked milling stone, its grinding surface downward *(6;* Fig. 37 *a).*

The photographs show the initial clearing of the burials in the adjacent sketch.

[52]

widely flaring rim (Fig. 9 *e*). One m north of the bule that accompanied Burial 7 and at same depth, but no evident reason to connect them with one another.

Burial 10 (Appendix V, no. 10305) from same pit as the bule of boneless Burial 8 and roughly 1 m east of the latter. This whole area produced small bits of bone and, moreover, ran into the undercut excavations of moneros. Otherwise, bone limited to a badly crushed skull and fragmentary mandible, plus a scapula and humerus. Mandible not articulated. Apparently burial had head to north, face to east. Evidently an excavation made for the corpse, a basin of Capacha monochrome (Fig. 12 *a*) placed upright in it, then the body introduced, head partly within the vessel. Basin rim 1 m below the surface.

"Oven"

In the main part of the cemetery, a certain amount of "informal excavation" had continued from 1970 to 1973. Adjacent to one monero pit was a large heap of medium-sized loose river cobbles, said to have been removed from an "oven." We decided to clear the monero pit and inspect what remained of the so-called oven, especially since we were told it had produced some charcoal. Among the stones still in situ, we collected a fair amount of charcoal (sample 250). The accompanying sherds were only 10, but all were Capacha monochrome. Lamentably, it turned out that the charcoal must have been contaminated; the resulting date is late in historic times (Table 1, no. 15).

This concentration of river boulders remains a mystery. The soil in contact with the stones was somewhat sooty; otherwise, except for the charcoal specimen, just mentioned, there was no evidence of fire; no ash, no burned earth, no stones that had obviously been subjected to high temperature. The cobbles had been concentrated between 50 and 100 cm below the surface; lateral dimensions not determinable owing to monero destruction. A tetrapod stone bowl allegedly found in the monero pit could not be located.

Comments

All burials at the Quintero cemetery that are identifiable as to phase are Capacha. However, the two large fragments of rose-red vessels found flush with the surface near Burial 6 are presumably relics of a more recent interment.

In the field, I was of the impression that—with exception of various griddle fragments—sherds from the surface and general digging were largely Capacha. Subsequently, when the material was washed and tabulated, a considerable lot of unclassified pottery showed up, especially from the vicinity of Burials 8 and 10. Included are Red and Unslipped red; Plain; and comal fragments. There are two sherds of Ortices shadow striped, none of other Ortices wares; none of Comala red; none of Amoles wiped. Quintero probably lies outside the range of Comala red, but certainly not that of Amoles wiped. Owing to the admixture, association of items from surface and general digging at Quintero is suspect; it may or may not be Capacha.

Several ceramic features warrant mention:

1. Our one complete stirrup pot with two tubes (Figs. 13 *d*; 25 *a, aa*); it presumably accompanied Burial 4 (Fig. 8, no. *3*).

2. A Capacha monochrome olla with complete rose wash; five repeats of an incomplete pendent triangle, with gouged fillers (Fig. 28 *a*).

3. Our only basin of Capacha monochrome (Fig. 12 *a*).

4. An entire Capacha figurine (Fig. 34 *c, cc, ccc*), hollow, recovered in the city of Colima long after its sale.

Appendix II
CERAMICS

This Appendix is essentially descriptive and includes a fairly detailed treatment of Capacha monochrome and its variants, as well as mention of unclassified ceramic material, including presumed trade wares. In addition, there is a brief discussion of griddles or comales, fragments of which occur at several Capacha cemeteries, perhaps out of chronological context. Miscellaneous pottery products are also described; of these, the rather spectacular figurines should be of particular interest.

CAPACHA MONOCHROME

Source Material

Ninety-seven vessels—entire, restorable, or semirestorable—of Capacha monochrome, plus a few slipped and painted variants. Cataloged sherds not abundant, for Capacha pottery occurs almost exclusively as burial furniture. Domestic wares may not be represented in present collection, despite the fact that most vessels look utilitarian. No soot-blackened cooking pots; few vessels suitable as receptacles for food or drink (cf. this appendix, *Unclassified and Trade Wares*).

Color

Cream to light brown, to gray, to dark brown, to black; all may occur on same vessel; occasionally bright orange from overfiring. Heavy firing clouds characteristic; isolated or sometimes covering half or more of vessel.

Thickness

Thick, generally 5 to 8 mm. Extreme range 3 to 14 mm, in one instance occurring on same specimen. Because of contour, thickness of some vessels (entire bules, certain ollas, and jars) can be measured only at neck and base.

Bules may have the base markedly thicker than the walls (Figs. 9 *b, j*; 15 *b*; 20 *a, b*). Sometimes the thickest part just above and below cincture (Fig. 9 *i*); occasionally, thinnest section at protuberance just above waist (Fig. 20 *e*). Ollas may have base thicker than body walls (Fig. 11 *b, d, g, j*). Jars and bowls show less disparity.

Paste

Somewhat grainy but not excessively so. In profile, dark streak common. Some vessels noticeably underfired; others, warped and bright orange from overfiring.

Slides of 11 sherds (prepared by courtesy of the Departamento de Prehistoria del Instituto Nacional de Antropología e Historia) were delivered to Ing. Adolphus

Langenscheidt for study. He reports (personal communication, January 1971): "From the mineralogical point of view, it may be said that all these sherds have temper which could be derived from local geological formations. Basically, the temper consists of particles of basalt and feldspar. Generally those of basalt are slightly rounded, which may indicate that they were intentionally ground."

Surface

Most vessels of large orifice (bules, ollas, bowls) have the interior quite well smoothed; smoothing marks sometimes visible. Interior is rarely polished. In contrast, exterior may be nicely smoothed and often has considerable polish, sometimes applied over a somewhat lumpy surface. Smoothing marks seldom noticeable. In some cases, surface has crackled and peeled. Specimens of restricted orifice (cántaros, stirrup pots) have a rough interior; exterior as above.

Technique

Manufacturing process uncertain; probably concentric rolls of clay superimposed. One sherd clearly indicates this procedure, and a bule (Fig. 9 *h*) has, on the interior of the lower body, a horizontal depression that suggests superimposed coils with junction not fully obliterated (this is not the exaggerated welt that results from horizontal mold common in much later times in Colima).

In cross section, an appreciable number of sherds have a crack parallel to the vessel wall. I was uncertain how to interpret this, but Dr. Wilhelm G. Solheim II suggests that: "Fracturing of pottery along the axis of the vessel wall can be the result of incomplete drying of the clay before firing. This [the crack] happens during the firing. I suspect that heavy polishing may give the same results both during firing and as a result of long-time weathering, but have not tested this" (personal communication, 24 September 1970).

Form

Capacha vessels at hand—exclusive of slipped and other painted variants—may be grouped as follows:

Bules: deep vessels, cinctured, with wide orifice
(Figs. 9, 15, 16, 17, 19, 20)   44
Cántaros (jars): deep vessels without cincture; restricted opening; suitable for liquids (Fig. 10 *a–g*)   10
Ollas (jars): deep vessels without cincture; wide aperture (Fig. 11 *a–j*)   11

Bowls (considerable variation): 19
    Basin (Fig. 12 *a*)   1
    Basin-cauldron (Fig. 12 *b*)   1
    Cauldron (Fig. 12 *c*)   4
    Hemispherical (Fig. 12 *d*)   1
    Incurved (Fig. 12 *e–k*)   9
    Flaring (Fig. 12 *l, m*)   2
    Plate (Fig. 12 *n*)   1
Stirrup pots and variants (Fig. 13 *d, e*)   3
Other compound vessels (Fig. 13 *a–c*)   3
Miniature vessels (Fig. 14)   6
Effigy pot (Fig. 13 *f*)   <u>1</u>
                         97

Bules

Almost half the Capacha vessels in the collection are bules (Figs. 9, 15, 16, 17, 19, 20), distinctive in form and ornament. Most characteristic aspect is the central constriction, from which the form derives its popular name, meaning "water gourd." The waist may be slight, either rounded or sharp, occasionally quite angular (Fig. 9 *k*). Range in size, form, and proportions shown in Figure 9. Moneros mention specimens with two, even three cinctures; I have seen one with a double waist, none with more. Sometimes the maximum diameter coincides with the body bulge above and below the constriction; more frequently, the lower body is of somewhat greater diameter. Very rarely does rim diameter exceed that of body (Fig. 9 *e*).

Usually the rim flares widely. When it does not (Fig. 9 *d, i, k*), the form suggests the belted vessels of the Tlatilco style of the central highlands, although with wider mouth; in no case does the Capacha bule approximate bottle form. As a rule, neck and rim are of uniform thickness, but sometimes there is a slight thickening on the underside of the rim (Fig. 9 *b*). Base usually rounded, and vessel cannot stand alone; a very few specimens are nearly flat-bottomed (Fig. 9 *j, k*), and one (Fig. 9 *e*) has on the exterior a slight depression in the center bottom.

One monero reported having found at Quintero a bule whose upper and lower bodies were separated on the interior by a clay divider, pierced by spaced perforations. He suggested, resourcefully, that the vessel might have been used for steaming medicinal preparations.[35] Unfortunately, this is the only indication of such a specimen.

Bule size range impressive. Smallest specimen at hand measures 7.8 cm in diameter at rim (Fig. 9 *j*); one sherd (not illustrated) is from a bule whose mouth was approximately 52 cm in diameter.

Cántaros

Cántaros (jars; Figs. 10 *a–g*; 27) are proportionately deep vessels, with neck and mouth relatively restricted. Occasionally difficult to distinguish between cántaros and ollas, the latter having proportionately wider orifice. Cántaro bodies vary considerably (Fig. 10 *a–g*). One with globular belly has elaborate incision and punctation (Figs. 24 *a*; 26 *a*); unfortunately, the neck is missing. Another cántaro has a tapering, subconic neck (Fig. 10 *d*). One vessel with similar neck and with red wash, of unknown provenience, is probably Capacha (Eisleb 1971: no. 192).

In the Colima area, this subconic neck occurs exclusively in Capacha association. From Tlapacoya, Weaver (1967: Fig. 5 *a*) illustrates a pot with similarly tapering neck. Description of a neck "slightly constricted towards the mouth" suggests such a form known also from Bajío times in the San Lorenzo Tenochtitlan series (Coe 1970: 24). I feel that these subconic necks do not quite qualify as bottles; the true bottle, with narrow, vertical neck, is extremely scarce in Capacha—if, indeed, it occurs. However, one sherd (Fig. 10 *h*) may have come from such a bottle.

Several cántaros and ollas have a faint break, even a slight channel, where neck and body meet (Figs. 10 *d, f, g*; 11 *h, i*). This is one specific and important resemblance between Capacha and Opeño ceramics (cf. Fig. 11 *k*, an Opeño specimen).

Ollas

Ollas (jars; Fig. 11 *a–j*) are comparatively deep pots, here called ollas in keeping with popular terminology in Colima. Wide mouth contrasts with constricted one of the cántaro.

Bowls

Bowls (Fig. 12) vary from an open basin (Fig. 12 *a*) and cauldron (Fig. 12 *c*) to simple hemispherical (Fig. 12 *d*) and incurved (Fig. 12 *e–k*) profiles. One plate (Fig. 12 *n*) possibly a pot cover, and a couple of small bowls with flaring rim (Fig. 12 *l, m*). All with rounded base.

An incurved bowl is the most common; underside of rim not thickened. A few rims pinched at opposite points (Fig. 12 *f, i*); viewed from above, vessel somewhat kidney-shaped. Several incurved bowls have perforations, usually paired, immediately below rim (Fig. 12 *f–h, k*), presumably for suspension.

Compound Vessels

Stirrup pots and variants (Figs. 13 *d, e*; 25) are vertically compound; of particular interest because of possible relationship with other areas (Chapter 4: *Tlatilco Style* and *Northwest South America*; Kelly 1972, 1974). Three specimens known to be from Colima; two (Fig. 25 *b, c*) were purchased, and the other (Fig. 25 *a, aa*) I excavated. There is no doubt of Capacha affiliation.

Of these three vessels, one is of the usual double-tube type (Figs. 13 *d*; 25 *a, aa*); the others have triple tubes (Figs. 13 *e*; 24 *c*; 25 *b, c*). In all cases, the tube profile is "broken" or "elbowed," as if jointed. Likewise, all three have the mouth in the form of a vessel, which, in the three-tube variants, is almost the size of the lower body. Aperture in the form of a vessel seems an almost unique Capacha feature; perhaps less so, the "elbowed" tubes. Finish varies: one specimen is black (Fig. 25 *a, aa*), another gray (Fig. 25 *c*), both with considerable polish. The third (Fig. 25 *b*) is brown and less polished. The gray specimen has grooved and punctate decoration.

In addition to these three stirrup vessels,[20] another three can presumably be assigned to the Capacha phase:

1. One, of unknown provenience, in the Museum of the American Indian, Heye Foundation (Fig. 25 *d*; Kelly 1972: Fig. 32 *d*; 1974: Fig. 1 *d*). Handsome. Polished black. Unmistakably Capacha in its trifid tubes and decoration.

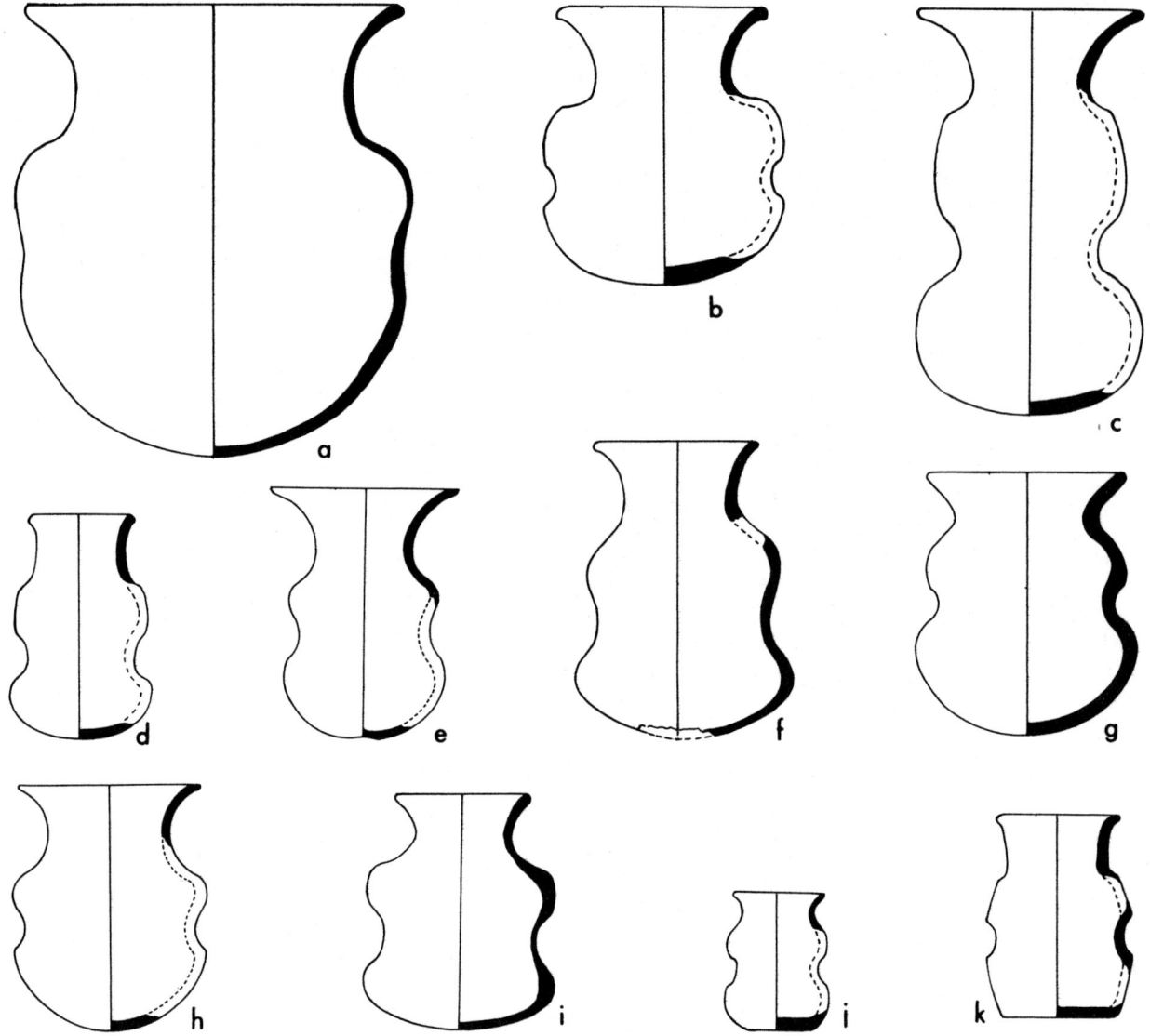

Fig. 9. Capacha monochrome. Vessel form: bules. See p. 24.
Scale 1/6. For *b, i, j,* see also p. 54; For *b. d, e, i–k,* p. 55; *j,* p. 47.

a. 5533. Appendix IV-E, unnumbered site, Potrero El Terrero. Purchased. See also Figs. 17 *a,* 19 *d.*
b. 9637. IV-D-2. See also Figs. 15 *b,* 17 *c.*
c. 8712. IV-C-6. See also Figs. 16 *f,* 17 *h.*
d. 8707. IV-A-6. See also Fig. 19 *c.*
e. 10303. IV-A-10. See also p. 53.
f. 10026. IV-E-10.
g. 9797. IV-B-2.
h. 9712. IV-A-2. See also p. 54; Fig. 4.
i. 8715 a. IV-E-6. See also Fig. 20 *f.*
j. 9651. IV-B-6. See also p. 55; Fig. 16 *d.*
k. 8684. IV-C-6. See also p. 55; Figs. 16 *e,* 17 *f.*

2. Another specimen (Fig. 25 *e*), likewise without known antecedents, but almost surely Capacha. It is in the Preclassic display in the Museo Diego Rivera-Anahuacalli, in Mexico City. I applied unsuccessfully for permission to photograph the vessel, which is of special interest in being the only (presumed) Capacha stirrup vessel I know whose orifice is not in the form of a well-defined pot. The specimen is small, dark gray, with two tubes and a short, straight spout, flaring slightly at the rim. Decoration is broad-line incision in the form of vertical stripes that contain opposed gashes.

3. The third vessel (Fig. 24 *c*) was photographed some 25 years ago in a private collection in Guadalajara, Jalisco. Notable in combining a ring or doughnut base with a typical three-tube stirrup mouth—a combination that might fit well with Dixon's suggestion that "the horizontal ring vessel and the stirrup-spout handle diffused together ... " (1964: 458). Other unusual features are the raised ridges on the upper body and on the ring. Despite such specialized traits, the specimen seems a Capacha product and of great potential significance in tracing ceramic relationships. It was reputedly found near Autlán, Jalisco, hence within the Armería drainage (Fig. 1). If provenience is reliable, this specimen indicates extension of the Capacha phase to the Autlán valley and thus links this area with Colima at a much earlier time level than heretofore supposed.

Miscellaneous compound vessels (Fig. 13 *a–c*) include a double bowl (Fig. 13 *b*); a monero described three united plates, which I did not see. Moreover, one fragment evidently formed part of a double jar (Fig. 13 *c*); another, of a triple jar (Fig. 13 *a*).

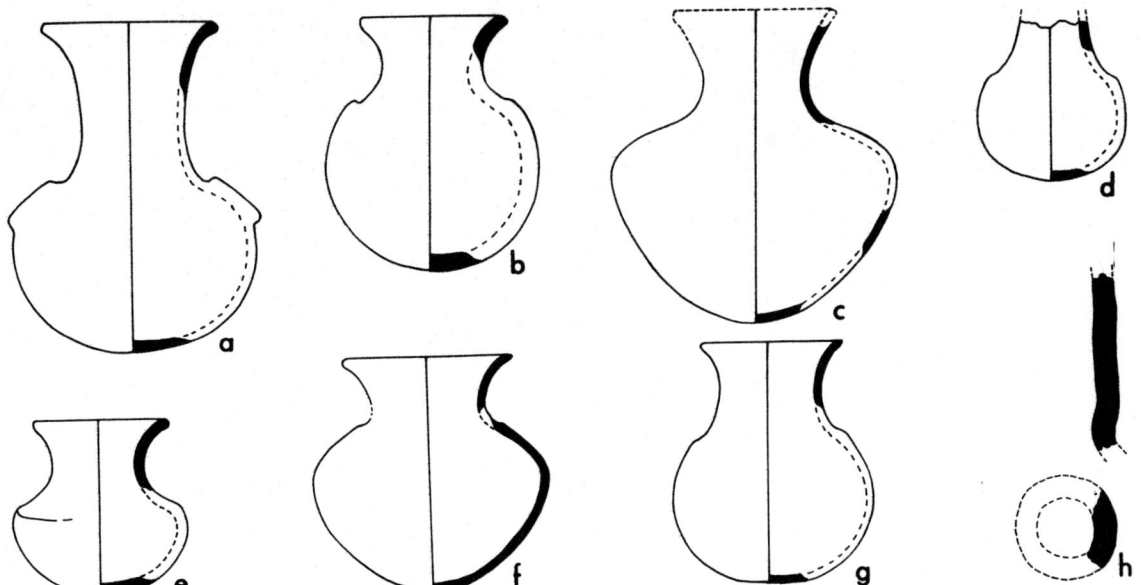

Fig. 10. Capacha monochrome. Vessel form: *a–g,* cántaros (although *d* might be considered a bottle variant with tapering neck); *h,* possible bottle-neck fragment. Scale 1/6. For *a–g* see also pp. 54, 55; *d, f, g, h,* pp. 26, 55; *d, f, g,* p. 39.

a. 8713. Appendix IV-C-6. See also Figs. 26 *c,* 27 *a.*
b. 9687. IV-E-10. See also Fig. 27 *b.*
c. 9634. IV-D-2.
d. 8888. IV-B-7. See also pp. 29, 34, 49; Fig. 11.

e. 10025. IV-E-10.
f. 9652. IV-B-6. See also p. 47; Figs. 11, 22 *a.*
g. 8679. IV-C-6.
h. 9917 a. IV-B-10. See also p. 25.

Miniature Vessels

Among the miniature vessels (Fig. 14) are one simple, open-mouth pot, poorly finished (*d*); an entire jar, with excrescences and with nubbin-tripod feet (*e*), and one fragment (*f*) from a specimen similar to the latter. Three tiny effigy vessels, tetrapod (*a–c*), were allegedly found in Capacha association. Footed supports occur only on miniature products.

Effigy Vessels

A bird effigy (Fig. 13 *f*), not a miniature, is definitely Capacha. Another specimen, from description similar, was reported from a Capacha cemetery but was not available for inspection. Still another,[20] apparently representing a tortoise, is included with the painted variants (Fig. 30 *c, cc*).

Decoration

Capacha ceramic ornament is dominated by broad-line incision or grooving; base of the depression is usually rounded in cross-section. Rarely is grooving so broad that it does not cut the surface (Figs. 17 *i;* 21 *a, e;* 29 *c*), but really sharp incision is infrequent (Fig. 20 *a, d*). Sometimes incision takes the form of hachure (Figs. 19 *b, c;* 20 *d, e*). What might be called "angular stripping" may be a variant form of incision; surface evidently cut with a fairly sharp, broad stylus, to judge from the sherd of one small olla (Fig. 31 *c*).

The most typical Capacha ceramic ornament is associated primarily with bules; found on both upper and lower bodies—that is, above and below the cincture. The characteristic design is a combination of broad-line incision and punctation, which I call a sunburst. Its center is a slight depression or "navel," evidently made by pressing

the moist clay with the fingertip. In several cases, this concavity clearly precedes the incision; probably it was the means of spacing the individual sunburst units. In a few specimens, the depression protrudes slightly on the interior vessel wall. From this central concavity, paired lines of broad incision extend at 45-degree angles, sometimes curved to follow the swell of the body. Such paired lines form four triangular fields, of which the upper and lower ones are filled with punches or gouges. Rarely do the lateral fields receive such treatment (Figs. 16 *e;* 17 *f*). Some instances of the basic sunburst without punctate fillers occur (Figs. 16 *d;* 17 *g, i*).

Incised delineation of the triangular fields clearly preceded punctation. Types of punctation quite varied. One bule and several sherds have circular depressions (Figs. 16 *b;* 18 *a, c*), diameters 2–5 mm. More commonly, the gouge elongated; tip of stylus may have been round, but insertion at a sharp angle resulted in an oval cavity. Occasionally, use of a sharp instrument yielded a narrow gash (Fig. 18 *i–k*). Sometimes form of gouge irregular and angular; occasionally, almost triangular (Fig. 18 *g*). In a few cases, a stylus with fibrous tip was used. Not always was the same instrument used to pit a given vessel.

Movement of the stylus not uniform. In some cases, upper body of bule jabbed with an upward thrust; lower body, with downward pressure; or the reverse. Some specimens have both bodies jabbed from below, with upward movement; in one case, the opposite is true. Clearly, motor habits were not uniform; they presumably varied with potter and with size and proportions of vessel.

As a rule, sunbursts uniformly spaced, generally with four repeats, sometimes six; rarely five, seven, or eight. Individual sunburst units generally connected by a pair of horizontal lines, which may open slightly to enclose an

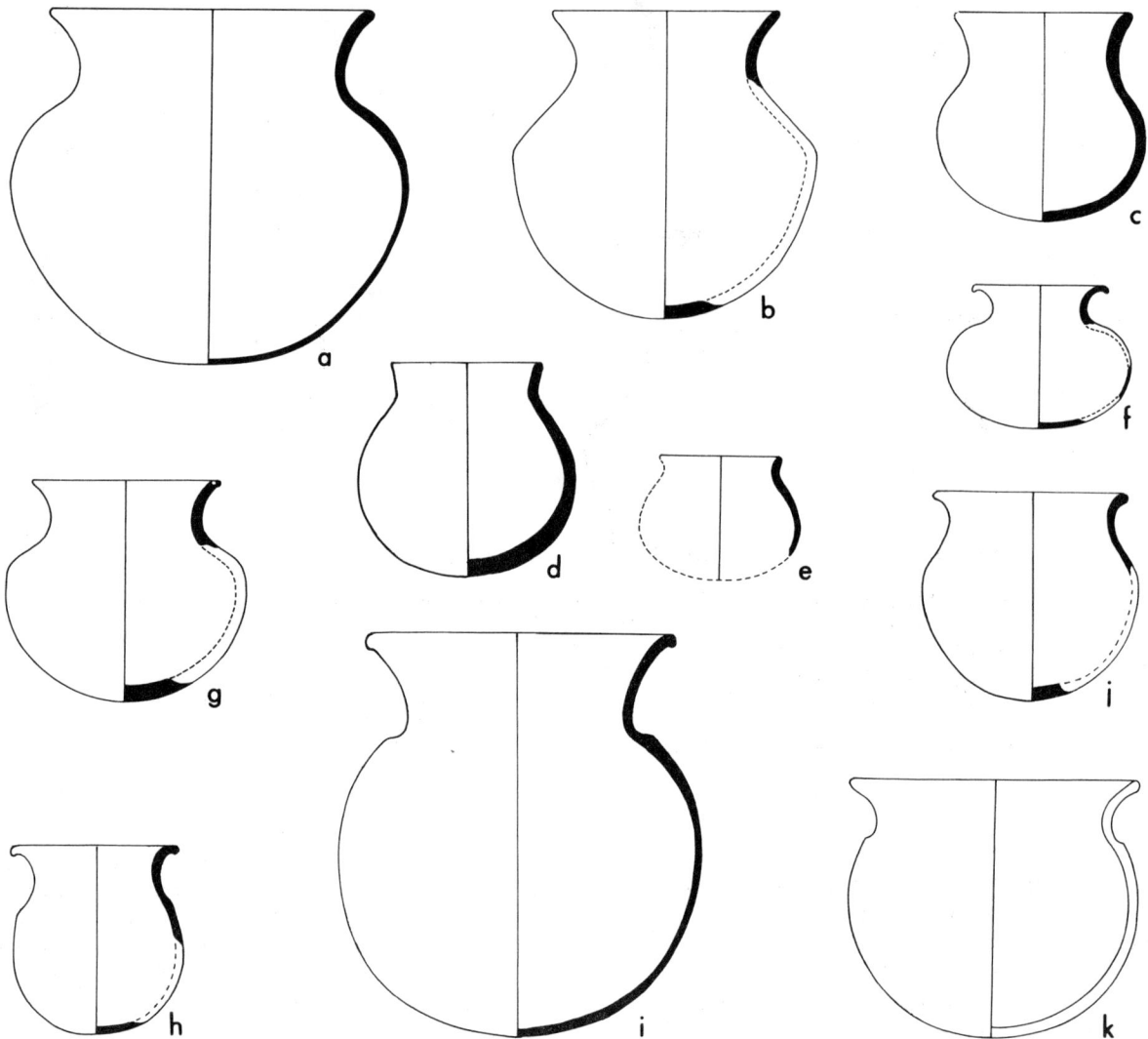

Fig. 11. Capacha monochrome and an Opeño vessel. Vessel form: ollas. *a–j*, Capacha monochrome, scale 1/6; *k*, olla from tomb at El Opeño, Michoacán; note close resemblance in form to *i* and particularly the break or faint channel below the throat, similar to that of *h*, *i* (cf. also Fig. 10 *d*, *f*). For *a–j* see also pp. 54, 55; *b*, *d*, *g*, *j*, p. 54; *h*, *i*, p. 39; *h*, *i*, *k*, p. 26.

a.	9639. Appendix IV-D-2.	*h.*	9711. IV-B-2.
b.	8677. IV-A-6.		See also p. 40.
	See also p. 18.	*i.*	9608. IV-A-3.
c.	9635. IV-D-2.		See also Fig. 6.
d.	9688. IV-E-10.	*j.*	9747. IV-A-10.
e.	10218. IV-B-8.		See also Figs. 8, 22 *b*, 23 *a*.
f.	9739. IV-B-10.	*k.*	After Oliveros 1970:
g.	9718. IV-A-2.		Fig. 22:64.
	See also Fig. 5.		See also pp. 30–31, 55.

ellipse; often the lines dip, garlandlike (Fig. 15 *d*, to note only one case). Rarely, a vertical panel (Figs. 16 *e*; 17 *f*) replaces the connecting lines. Sometimes individual sunbursts are contiguous (Figs. 15 *c*; 16 *f*; 17 *g–i*; 20 *c*); on one jar sherd they are continuous and constitute an allover network or lattice pattern (Fig. 24 *a*).

Table 3 gives the distribution of ornament among the 44 bules. Clearly, the sunburst is the most popular decoration. It sometimes occurs above the cincture, with the lower body undecorated. More frequently, a full sunburst on both bodies, or a complete one above, an incomplete one below. In latter case, upper half of sunburst is the part usually depicted on lower body (Figs. 16 *a*; 19 *a*).

Although characteristic of bule decoration, the sunburst is not limited to that vessel form. A cántaro fragment (Figs. 24 *a*; 26 *a*) has it worked into a lattice or network which, on the base of the vessel, terminates as a seven-point star. Otherwise, only one cántaro (Fig. 22 *a*) is incised. In contrast, two stirrup pots have sunburst ornament (Fig. 25 *c*, *d*), but another (Fig. 25 *e*) has simple opposed gouges contained within a band formed by two horizontally incised lines.

There must have been a much wider range of decoration than the foregoing suggests.[20] Ollas and bowls (Figs. 22 *b*, *c*, *e*, *f*; 23 *a–c*) have incised and punched ornament, less elaborate than the sunburst. A selection appears in Figure

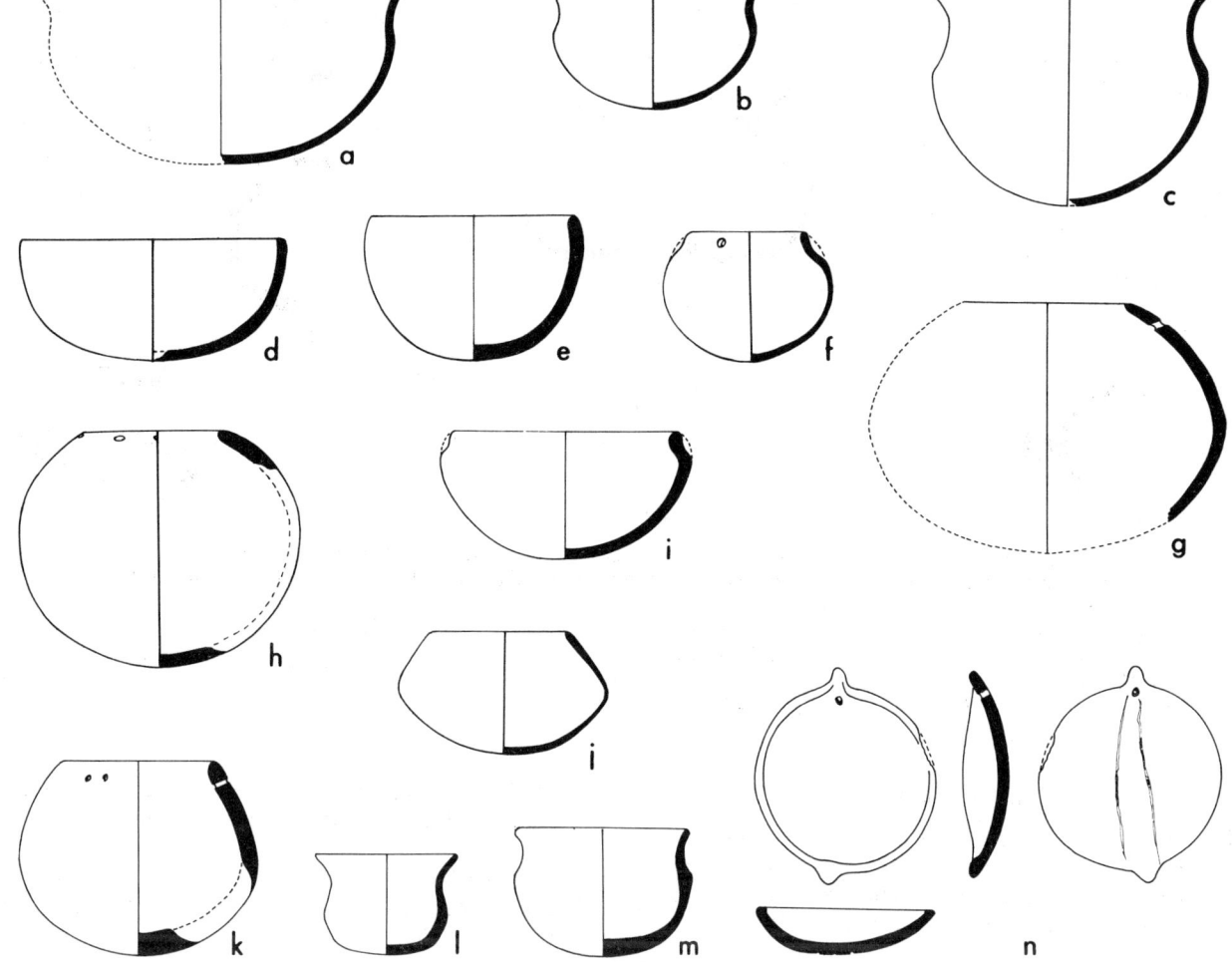

Fig. 12. Capacha monochrome. Vessel form: bowls. *a,* basin; *b,* intermediate between basin and cauldron; *c,* cauldron; *d,* hemispherical (the only example of this form); *e–k,* incurved (*e,* very slight incurve at rim; *f, i,* pinched rim; *f, g, h, k,* with rim perforations; *j,* a lone example with carination); *l, m,* flaring, both vessels unique in form and in lumpy, unpolished surface; *n,* plate with excrescences and perforation, possibly a pot cover (suggestion of Profa. Rosa María Reyna, who points out the feasibility of tying such a lid in place by means of corresponding rim perforations on vessel and on cover). Scale 1/6. For *a, c–n,* see p. 55; *f–h, k,* pp. 24, 55.

a.	10304. Appendix IV-A-10. See also p. 53.	*h.*	9749. IV-B-10. See also Figs. 22 *f,* 23 *b.*
b.	8716. IV-E-6.	*i.*	9685. IV-E-10. See also pp. 24, 35, 59.
c.	10242. IV-A-5. See also p. 46.	*j.*	10022. IV-E-10.
d.	10021. IV-E-10. See also Fig. 22 *d.*	*k.*	10023. IV-E-10.
e.	9684. IV-E-10. See also p. 40.	*l.*	9633 a. IV-D-2. See also p. 40.
f.	9686. IV-E-10. See also pp. 35, 59.	*m.*	9636. IV-D-2.
g.	9971. IV-E-8.	*n.*	10019. IV-E-10. See also p. 59.

31 *a–d, h–j.* One sherd has a stripped surface (Fig. 31 *c*) and in one case, the incision has a punctate start (Fig. 31 *d*), a trait reminiscent of northwest South America and of eastern Colima in more recent phases. Other examples are unusual in their round punctation (Fig. 31 *b, d, h–j*). Figure 31 *i* is almost paper-thin, uncommon for Capacha. Figure 31 *j* is from some sort of beaker or tumbler, not otherwise represented in our Capacha sample; it also is thin by Capacha standards, and the polishing striations are atypical.

On the chance that trade wares might be involved, tiny fragments of two of the atypical sherds (Fig. 31 *h* and one other, not illustrated) were studied by Dr. Harbottle, and a sample of the sherd shown in Figure 31 *j* was inspected by Ing. Langenscheidt. In neither case was nonlocal origin indicated.

In addition to incision and punctation, Capacha monochrome has a certain amount of modeling.[20] Ribbing or ridging may be so slight as to resemble faint finger grooving, or it may be exaggerated (Figs. 26 *b;* 27 *a;* 31 *b*). The ridging may be at right angles (Fig. 26 *d*) or disposed more elaborately (Eisleb 1971: no. 192). The Eisleb specimen just cited is probably a Capacha variant with rose slip, as are two ribbed sherds (Fig. 29 *h, i*) in the Colima collection. A trifid-spout vessel with doughnut base, presumably Capacha, has short, raised ribs in sets of three, on upper and lower bodies (Fig. 24 *c*). Protuberances on the rim of our one double bowl (Fig. 26 *e*) and of a simple plate (Fig. 12 *n*) may be considered forms of modeling, as are the pinched rims of bowls (Fig. 12 *f, i*). Effigy vessels—both monochrome and painted variants—demonstrate Capacha skill in animal portrayal (Figs. 26 *f;* 30 *c, cc*).[20]

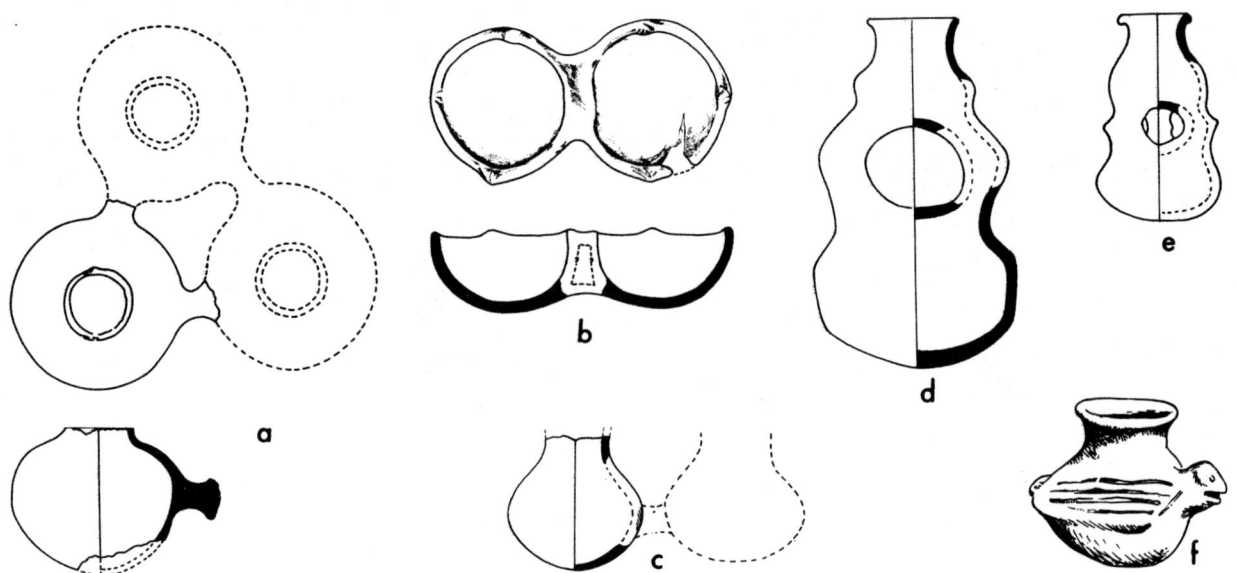

Fig. 13. Capacha monochrome. Vessel form: *a–e*, compound vessels; *f*, effigy pot. *a–c*, horizontally compound; *d*, *e*, vertically compound (stirrup pot and three-tube variant). Scale 1/6. For *a–e*, see also pp. 25, 56; *a–f*, p. 55; *b*, p. 57; *d*, p. 35.

a. 10024. Appendix IV-E-10.
    See also p. 26.
b. 9632. IV-D-2.
    See also p. 42; Fig. 26 *e*.

c. 8717. IV-E-6.
d. 9743. IV-A-10.
    See also pp. 19, 26, 35, 53, 55;
    Figs. 8, 25 *a, aa*.

e. 8714. IV-C-6.
    See also pp. 26, 55; Fig. 25 *b*.
f. 8706. IV-A-6.
    See also pp. 18, 26, 49, 57; Fig. 26 *f*.

Painted Variants

Painted wares—red slipped, red on brown, and a purple-black and red combination—are probably more frequent than present data suggest. There are so few examples that most are described individually below.

Red Slipped

A medium-sized olla (Figs. 28 *a*; 29 *b*) with rose-red wash on throat and entire exterior, except base. Thickness, paste, shape, as well as incised and punched decoration, typically Capacha. Interior nicely smoothed; exterior, lumpy and uneven. Base and lower wall a single disk of clay, to which were added concentric(?) circles to form body and neck. Slip preceded incision and punctation. At base of neck, a single line of broad incision from which are pendent five ill-defined and incomplete triangles, likewise in broad-line incision; these filled with sharp, irregular jabs or punches.

In addition, 12 sherds are rose slipped (Fig. 29 *c–i*), including two olla fragments with ridged relief (Fig. 29 *h, i*).

Red on Brown

An olla (Figs. 28 *c*; 30 *a*) has rose throat and neck; on exterior, same color extends from rim to faint channel, below which are six pendent rose stripes.

A second olla (not illustrated; Appendix IV-E-2, 9641 *a*) has a globular body and low neck; four slanting stripes pendent from the latter are rosy maroon. Possibly Capacha, although sharp neck-body junction atypical. In addition, one red on brown sherd (Fig. 31 *f*) is from an olla rim.

Unusual vessel form, best classed as bowl with flaring rim (Fig. 28 *d*); almost surely Capacha. Rose rim and apparently five pairs of vertical stripes on exterior.

Bowl sherd (Figs. 28 *e*; 31 *e*), with pinched rim and rim perforation. Exterior with unzoned red on brown decoration.

Incurved bowl, with four unzoned rose triangles pendent from rim, their bases alternating with pairs of rim perforations (Fig. 28 *f*).

Red on Brown, Incised Outlining

Neck, cincture, triangles on upper body and a narrow band on lower belly are a light rose (Fig. 30 *d*). This is the only painted bule I know. Incision and punctation filled with white substance, perhaps lime. Typologically, specimen cannot be other than Capacha, although antecedents unknown.

Restorable olla (Fig. 30 *b, bb*), with rose-maroon rim, neck, and body design. Latter delineated by broad-line incision applied subsequent to painting. Punctation in form of sharp jabs made with thin instrument. As in preceding case, white material in interstices.

Extraordinary effigy vessel apparently representing a tortoise (Fig. 30 *c, cc*). Upper part of pot represents the scute; head, tail, and feet are shown in relief, as is spinal column, the latter with sharp, opposed diagonal slashes. Relief combined with red on brown diamonds outlined by relatively broad-line incision, applied after painting. White material in interstices.

Red and Black, Incised Outlining

Painted ornament not confined to red on brown; moneros mention bules painted in red and black from El Barrigón (Fig. 1, no. *7*) and Parcela de Luis Salazar (Fig. 1, no. *5*). I have not seen such vessels, but two sherds from the surface of El Barrigón (one, Fig. 31 *g*) answer moneros' description. Paste is coarse, grainy, typically Capacha. Interior of vessel is light brown, almost cream; nicely finished, indicating bule or wide-mouthed olla. Exterior of

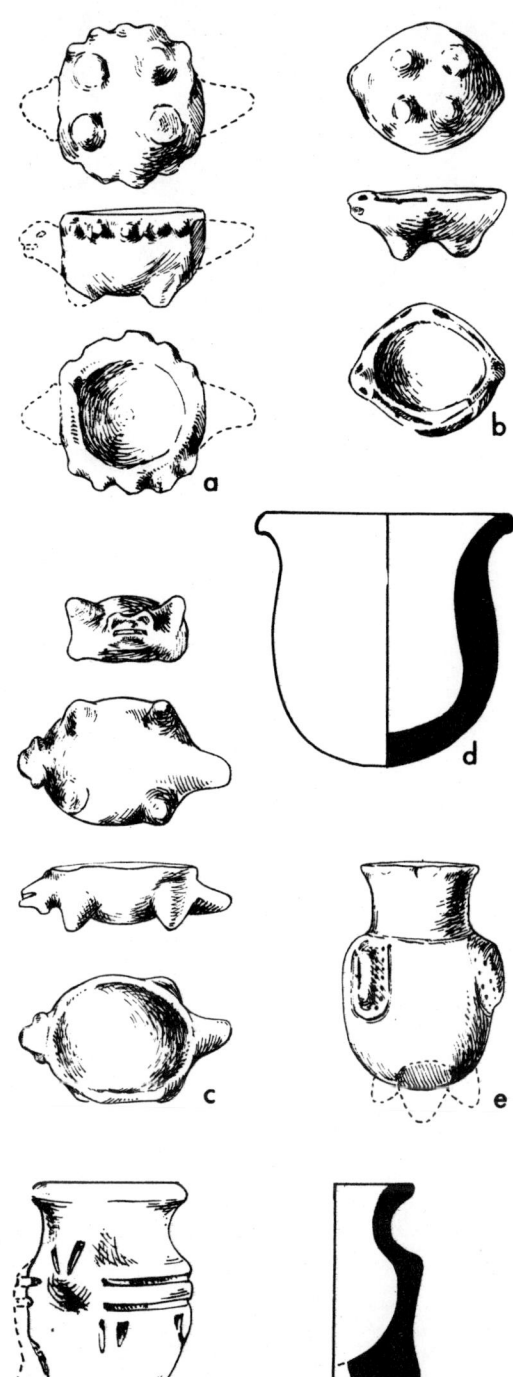

Fig. 14. Capacha monochrome. Vessel form: miniatures. *a–c,* Animal effigies, tetrapod; *d,* excessively heavy and crude; *e, f,* tripod (*e,* with three excrescences, plus grooving and punctation; *f,* grooving and coarse gashes). Scale ½. See also p. 55. For *a–c,* see also pp. 26, Note 26 (p. 103); *a–c, e, f,* p. 26; *a–f,* p. 57.

a. 8685 a. Appendix IV-C-6.
b. 8685 b. IV-C-6.
c. 8710. IV-C-6.
d. 10243. IV-A-5. See also p. 46.
e. 9621. IV-B-3. See also p. 44.
f. 10216 a. IV-B-10.

with purple-black geometric ornament outlined by incision, which is filled with white material (cf. Chapter 4: *Opeño Phase*).

UNCLASSIFIED AND TRADE WARES

With a certain frequency, unclassified Red and Plain sherds (Fig. 31 *l,* unusual in its lumpy surface) come from the surface and general digging at Capacha cemeteries, the Terreno de Fidel Valladares (Fig. 1, no. *8*) being an exception (Appendix IV-B-8). Elsewhere, such unclassified material is quite abundant, and in the vicinity of Burials 8 and 10 at Quintero (Fig. 1, no. *10*), such sherds actually outnumber those of Capacha monochrome.

Obviously, a cemetery churned by sequential reuse, as well as by devastating monero depredations, may, with luck, produce a few reliable grave lots, but otherwise it is not very helpful chronologically. I personally removed every Capacha-phase vessel we found, to check for underlying material. At La Parranda A (Fig. 1, no. *3*), a small sherd of unslipped Red unclassified was found directly beneath the olla of Capacha monochrome that accompanied Burial 3, and at the same cemetery a similar fragment was the only sherd to come from the subsoil pit that contained Capacha-phase Burial 2. There is, then, the possibility that an unslipped Red unclassified ware may be contemporaneous with Capacha monochrome, or even earlier. It is possible that the Capacha monochrome was made for burial furniture, while the puzzling Red and Plain unclassified wares were for domestic use. These latter have not been studied closely but seem a motley lot. Vessel form is largely indeterminate owing to the scarcity of rim sherds, but there is no evident resemblance to Capacha monochrome in paste, surface finish, or shape. Nor, at present, is there any instance in which such an unclassified vessel accompanies a Capacha burial, unless the pot described below is such an example:

The vessel in question (Figs. 28 *b;* 29 *a*) comes allegedly from a specific monero pit at La Cañada. It fits well with Capacha material in form and in decorative finger grooving, but its paste is fine, with a wide central core of dark gray, flanked at both surfaces by rosy red. Latter color evident on entire interior. Exterior, from rim to an almost imperceptible shoulder, also rose red, but apparently from a partial slip. Below, the belly seems natural color, except where the rose wash continues downward in the depressions formed by five pairs of finger grooves. This appears a deliberate application of rose paint, making the pot tech-

one sherd painted red and black—not black on red, although in some spots the black overlaps the red. Subsequent to painting, the two colors divided by broad-line incision. The red is a soft rose; the black has strong purplish cast and a noticeable sparkle, which Ing. Adolphus Langenscheidt attributes to particles of specular hematite. The second sherd (not illustrated), probably once very similar, perhaps even from the same vessel; now badly weathered. Under hand glass, its "black" looks dark reddish purple.

An additional red and black specimen, a cántaro, clearly related to Capacha, comes from near Tuxcacuesco (Fig. 30 *e;* Kelly 1949: 83–84, Pl. 14 *d*). It is polished red,

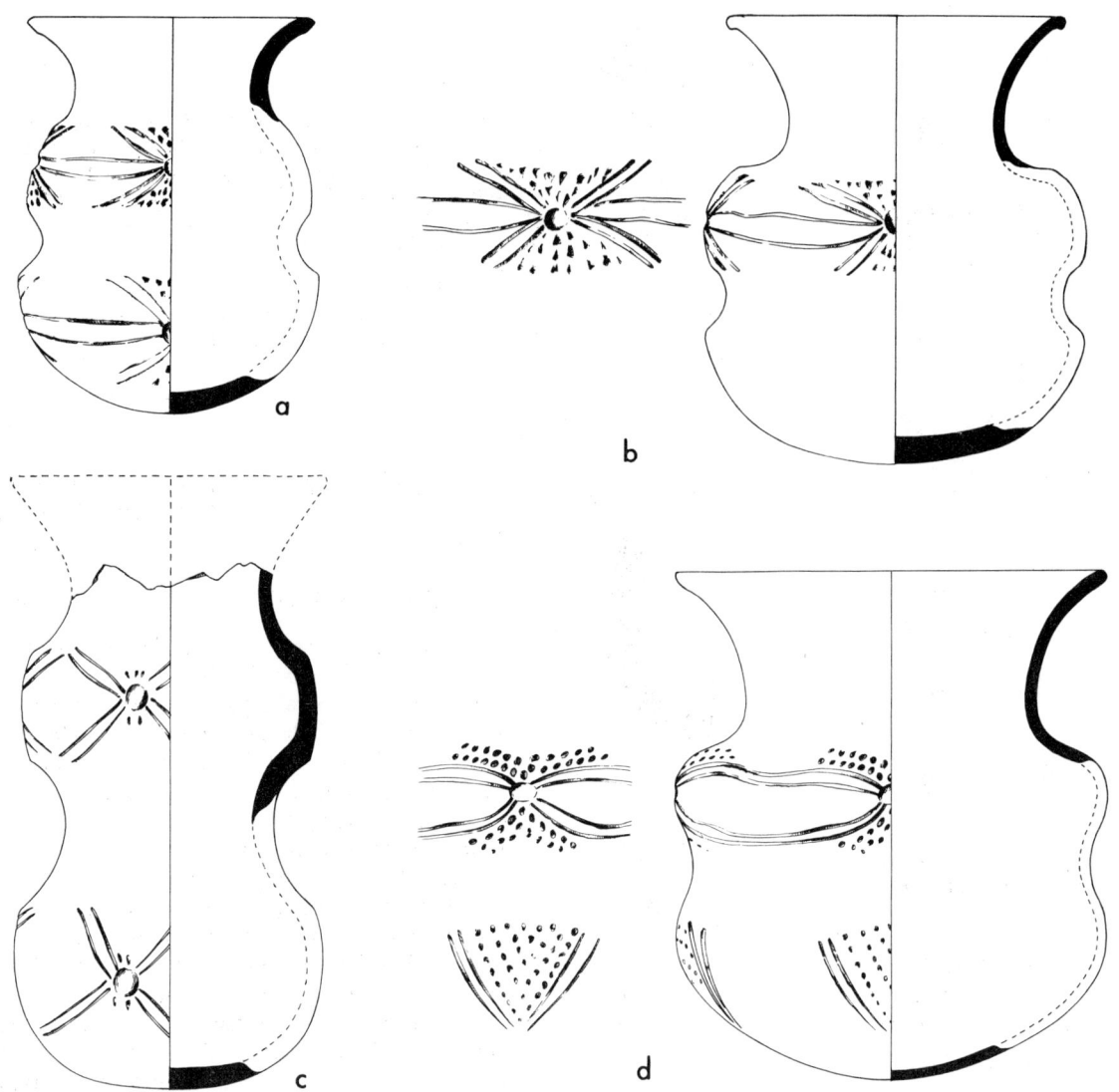

Fig. 15. Capacha monochrome. Decoration: incision and punctation; sunburst motif. Bules; all with four repeats. Scale ¼. See pp. 26, 54, 55, 66 (Table 3); for *c*, *d*, see also p. 58.

a. 9713. Appendix IV-A-2. See also Figs. 4, 17 *b.*

b. 9637. IV-D-2. See also Figs. 9 *b*, 17 *c.*

c. 9690. IV-E-10.

d. 9609. IV-A-3. See also Figs. 7, 17 *d.*

nically red on brown. Except for shape and finger grooving, specimen seems quite different from Capacha and could relate to some of the Red unclassified that crops up at Capacha cemeteries. It is a pity we do not have full information concerning its appearance at La Cañada.

Two sherds with excised ornament must be considered trade products; excision is not a local trait at any time level and strongly suggests contact with the Mesoamerican Preclassic. Neither fragment comes in absolutely sure Capacha association. The first (Fig. 31 *k*) has a grainy paste, but fine by Capacha standards; tiny particles (mica?) give a surface sheen. Apparently from an open bowl; interior brown to black, nicely smoothed. Exterior nearly black; excised areas extremely rough; no trace of post-firing pigment. Design curvilinear, with a couple of incised lines in addition to excision. The second excised sherd (not illustrated; Appendix IV-B-8, 10219 a) is from the surface of a site whose sherds are chiefly Capacha. It is small, thin;

may come from a bowl. Paste fine by Capacha standards. Excised design vaguely curvilinear; cut into an area that has brownish-red slip. No trace of post-firing pigment.

GRIDDLES
(NON-CAPACHA?)

The occurrence of comal or griddle sherds is puzzling. Although paste, color, and texture certainly do not suggest the Capacha phase, a considerable number of fragments come from Capacha cemeteries, in all cases, associated with mixed sherd lots. Incidence is:

La Parranda A, pit 6	2
La Parranda A, overlying Burial 3	2
Parcela de Luis Salazar	1
Quintero (Fig. 31 *m*)	55

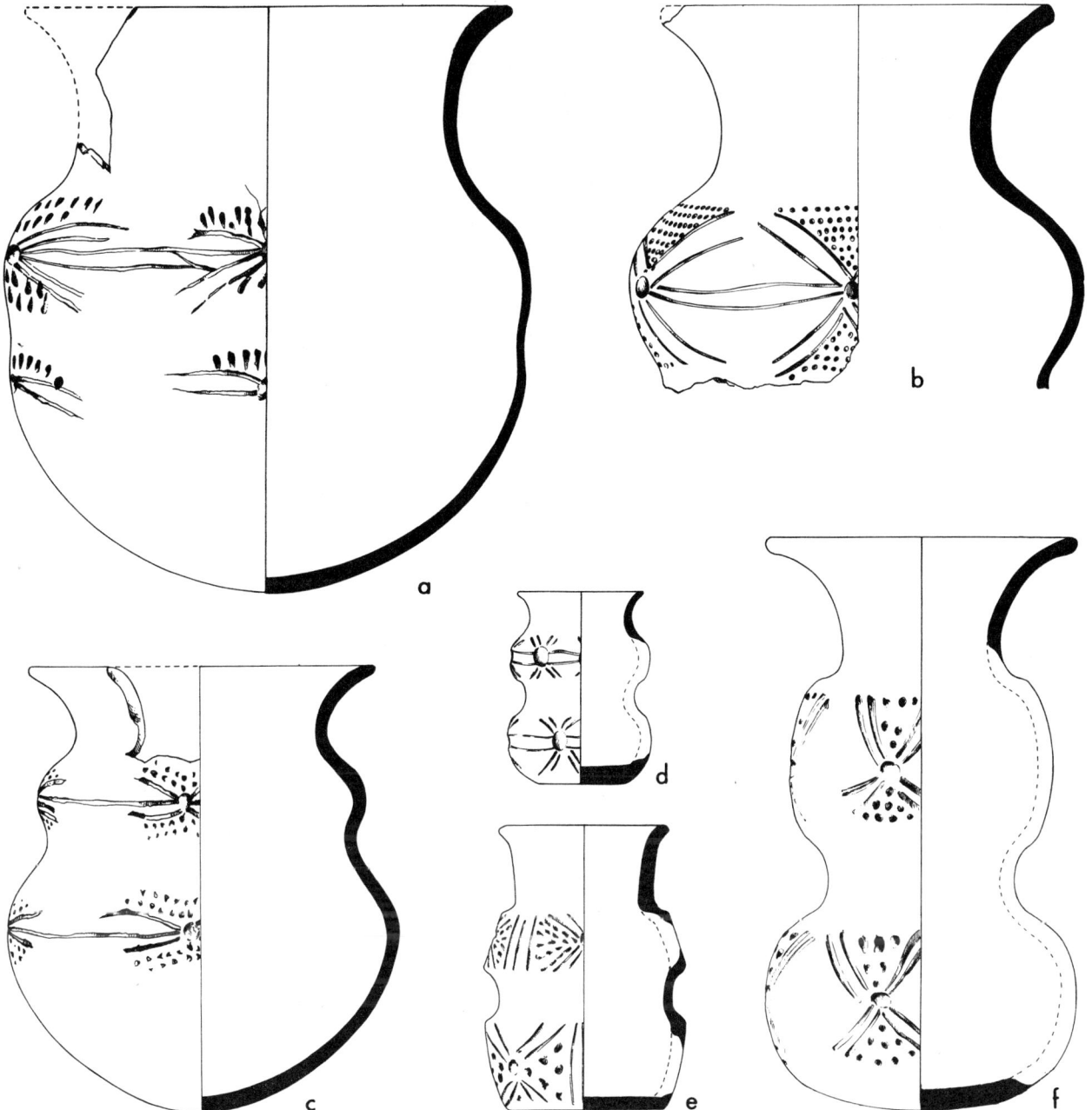

Fig. 16. Capacha monochrome. Decoration: incision and punctation; sunburst motif. Bules. Scale ¼.
See pp. 26, 54, 55, 66 (Table 3). For *b*, see also p. 57; *a, e, f,* p. 58.

a. 9607. Appendix IV-A-3. Four repeats, upper
    and lower bodies. See also p. 58; Fig. 6.

b. 9803. IV-B-2. Large sherd, possibly five re-
    peats. See also Fig. 18 *a*.

c. 9798. IV-B-2. Four repeats, upper and lower
    bodies. See also Fig. 18 *e*.

d. 9651. IV-B-6. Seven repeats, upper body; six,
    lower body. See also pp. 47, 57; Fig. 9 *j*.

e. 8684. IV-C-6. Four repeats. See also pp. 18,
    57; Figs. 9 *k*, 17 *f*.

f. 8712. IV-C-6. Six repeats, upper and lower
    bodies. See also Figs. 9 *c*, 17 *h*.

The impressive number of fragments from Quintero is explained by the fact that it has a varied sherd mixture, including a considerable ingredient of Red and Plain unclassified, especially from the vicinity of Burials 8 and 10.

The comal·sherds have a reddish, quite fine paste; the unslipped surface is a soft, reddish maroon-brown. The floor is nicely smoothed; to some extent, the rim is smoothed, but the exterior below it is rough and spongy (Fig. 31 *m*). Pottery griddles in modern rural Mexico have this same rough undersurface; in fact, the modern Totonac form the comal on a bed of sand or ash, which contributes to the rough quality (Kelly and Palerm 1952: 218). The illustrated sherd is perforated near the rim, thus forming a loop for suspension.

Fig. 17. Capacha monochrome. Decoration: *a–f, h,* incision and punctation; *g, i,* broad-line incision. Sunburst motif. Bules. Not to scale; size indicated below. See pp. 26, 54, 55, 66 (Table 3).

a. 5533. Appendix IV-E, unnumbered site, Potrero El Terrero (p. 96). Rim diameter, 32.5 cm. See also p. 18; Figs. 9 *a,* 19 *d.*

b. 9713. IV-A-2. Exterior rim diameter, 14 cm. See also Figs. 4, 15 *a.*

c. 9637. IV-D-2. Rim diameter, 18 cm. See also Figs. 9 *b,* 15 *b.*

d. 9609. IV-A-3. Rim diameter, 23.0–23.7 cm. See also Figs. 7, 15 *d.*

e. 9719. IV-A-2. Rim diameter, 26.5–27.0 cm. See also p. 24; Fig. 5.

f. 8684. IV-C-6. Rim diameter, 11 cm. See also pp. 18, 57, 58; Figs. 9 *k;* 16 *e.*

g. 10027. IV-E-10. Rim diameter, 15 cm. See pp. 57, 58.

h. 8712. IV-C-6. Rim diameter, 19.5–20.0 cm. See also pp. 33, 58; Figs. 9 *c,* 16 *f.*

i. 10302. IV-A-10. Rim diameter, 15.5 cm. See also pp. 51, 57, 58.

Fig. 18. Capacha monochrome. Decoration: incision and punctation; variants of the sunburst. Note the small, circular punches (*a*), ranging to longitudinal gashes (*i–k*), and the incision, which runs from broad-line grooving (*b, c, f, h,*) to relatively sharp, narrow cuts (*i*). Bule sherds. With exception of *a* and *e,* not in Appendix IV. Scale not uniform. For *a, b, d–k,* see also p. 26; *a, c, g, i–k,* p. 57.

a. 9803. Appendix IV-B-2.
    See also p. 31; Fig. 16 *b.*
b. 9837 b.
c. 10224 b.
    See also p. 31.

d. 9805 c.
e. 9798. Appendix IV-B-2.
    See also Fig. 16 *c.*
f. 8192 t.
g. 9913 a.

h. 9960 b.
i. 9252 a.
j. 8192 m.
k. 9894 a.

Inasmuch as a flat griddle—although of somewhat different style—evidently is late in Colima (see Chapter 2: *Chanal; Periquillo*), this apparently early occurrence is surprising. It is tempting to think that the comal in question, along with the Red and Plain unclassified that crop up at Capacha cemeteries, may belong to the domestic wares of that phase.

Elsewhere, there are scattered occurrences of griddle-like plates in "early" context—for example, in Honduras (Coe 1961: 126–27, citing Canby), and at Momil (G. and A. Reichel-Dolmatoff 1956: 270–71). The possibility that these comales may have been used in the preparation of bitter manioc has been discussed at length (G. and A. Reichel-Dolmatoff 1956: 270–72; G. Reichel-Dolmatoff

TABLE 3

Capacha Monochrome: Bule Decoration

Upper Body	Lower Body	Number of Cases
1. Undecorated	Undecorated (Fig. 9e–h, and three others not illustrated)	7
2. Horizontal lines, continuous and broken	Horizontal lines, continuous and broken (Fig. 20a)	1
3. Primarily vertical hatch	Primarily chevron hatch (Fig. 20d)	1
4. Vertical hatch	Diagonal hatch (Figs. 9d, 19c—same specimen)	1
5. Diagonal hatch	Diagonal hatch (Fig. 20b)	1
6. Diagonal hatch, units at opposed angles (approaching chevron hatch)	Undecorated (Figs. 9i, 20f—same specimen)	1
7. Cross hatch	Diagonal hatch (Fig. 20e)	1
8. Cross hatch	Cross hatch (Fig. 19b)	1
9. Paired lines, vertical hatch; punches in vertical row	Sunburst (Fig. 20c)	1
10. Incomplete sunburst	Undecorated (not illustrated)	1
11. Sunburst	Undecorated (Figs. 9b, 15b, 17c—same specimen; and five others not illustrated)	6
12. Sunburst	Sunburst (Figs. 9c, 16f, 17h—same specimen; 9j, 16d—same specimen; 9k, 16e, 17f—same specimen; 15a, 17b—same specimen; 15c; 16c, 18e—same specimen; 17g; and three others not illustrated)	10
13. Sunburst	Incomplete sunburst (Figs. 16a; 17e; 19a; and five others not illustrated)	8
14. Sunburst	Indeterminate, vessel incomplete (Figs. 16b, 18a—same specimen)	1
15. Sunburst	Chevron, punctate filler; close to an incomplete sunburst (Figs. 15d, 17d—same specimen)	1
16. Sunburst	Arc surrounded by punches (Figs. 9a, 17a, 19d—same specimen)	1
17. Sunburst variant, broad grooving, no punctation	Sunburst variant, broad grooving, no punctation (Fig. 17i)	1
		44

Note: Tabulation is based on 44 specimens that are whole, restorable, or nearly restorable. References to Figure 9 are included, although that figure shows only bule form. As indicated, in various instances, a given specimen is illustrated more than once, in order to demonstrate form, profile, decoration, and texture. Overwhelming preference for the sunburst decoration is evident.

1957: 233; Evans and Meggers 1960: 341–43; Lowe, in Green and Lowe 1967: 58–60). Evans and Meggers point out that an absence of pottery griddles need not imply absence of bitter manioc.

Nor, it may be added, does such absence necessarily reflect absence of maize. An old source (Relación de Ameca 1878: 265–66) reports that in one area of Jalisco, at the time of the Spanish Conquest, maize was eaten as tamales, *poleadas,* and toasted grains (*cacalotl* or *izquitl*). The tamal needs little explanation; it consists of maize dough, usually with a filling that often is of meat, enclosed in corn husks (in some areas, in banana leaf), and steamed. A sixteenth-century account (Cervantes de Salazar 1914, I: 17) states that atole (maize gruel) is similar to "poleadas of Castillà." For atole, Cervantes describes the maize as toasted on a comal, then ground and prepared as gruel. Nevertheless, for none of the three dishes is a comal-griddle really essential. The maize could have been wet-ground on a metate or pounded in a mortar and the resulting dough either steamed to make tamales or boiled to prepare the atolelike poleadas. Furthermore, dry maize kernels could have been toasted in an olla, as is sometimes done today in rural areas of west Mexico. In the Ameca account, "thin tortillas" apparently are a post-Conquest innovation.

Furthermore, my resourceful housekeeper, Sra. Aurelia Contreras—unaware of the august company mentioned above—inspected the griddle sherds from Quintero and said immediately that they were similar to the comales her grandmother had used to toast seeds such as cacao, squash, and chili (*chile ancho, pasilla,* and *mulato*). In other words, the comal—early or late—is no guarantee of the presence of maize in the diet, for it may be no more than a ceramic equivalent of the basket tray used by Great Basin and California Indians for processing wild seeds. By the same token, Evans and Meggers (1960: 341) point out that the presence of milling stones does not necessarily imply cultivation of maize.

Whatever the interpretation, the chronological position of the comal in the Colima area is unclear. If the fragments that crop up in the Capacha cemeteries prove to be "early," then temporal discontinuity is marked.

MISCELLANEOUS POTTERY PRODUCTS

Figurines

Several figurines are attributable to the Capacha phase (see Chapter 3: *Miscellaneous Manufactures*). They tend to be underfired, and moneros who dug at La Cañada referred repeatedly to *monos crudos.* To judge from the fragments they left discarded on the surface, the expression is apt.

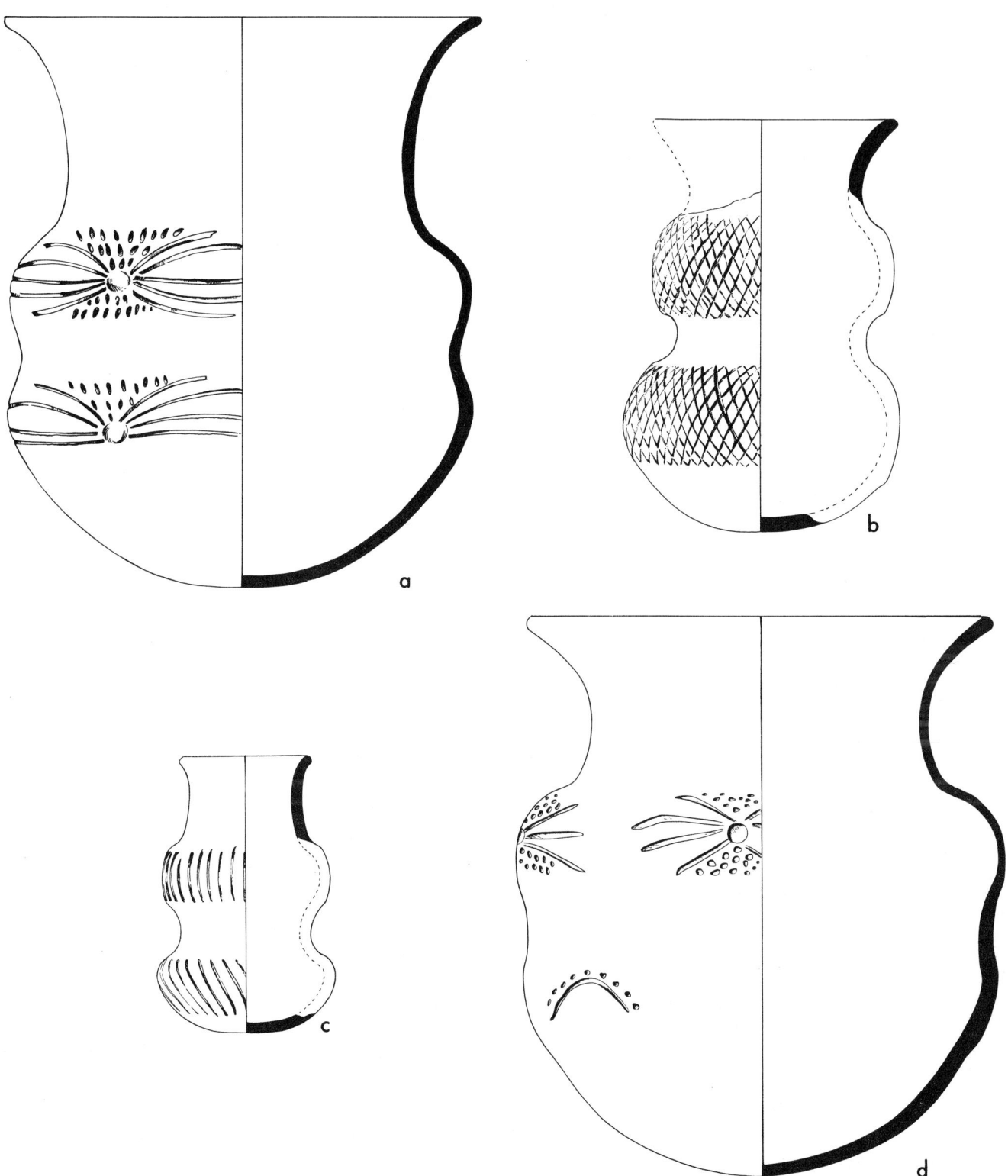

Fig. 19. Capacha monochrome. Decoration: incision and punctation; various motifs. Bules. Scale 1/4. See pp. 54, 55, 66 (Table 3).

a. 9605. Appendix IV-A-3. Fragmentary; number of repeats uncertain. See also pp. 26, 58.
b. 8633. IV-A-7. See also pp. 18, 48, 57.
c. 8707. IV-A-6. See also pp. 18. 57; Fig. 9 d.
d. 5533. IV-E, unnumbered site, Potrero El Terrero (p. 96). See also pp. 18, 26; Figs. 9 a, 17 a.

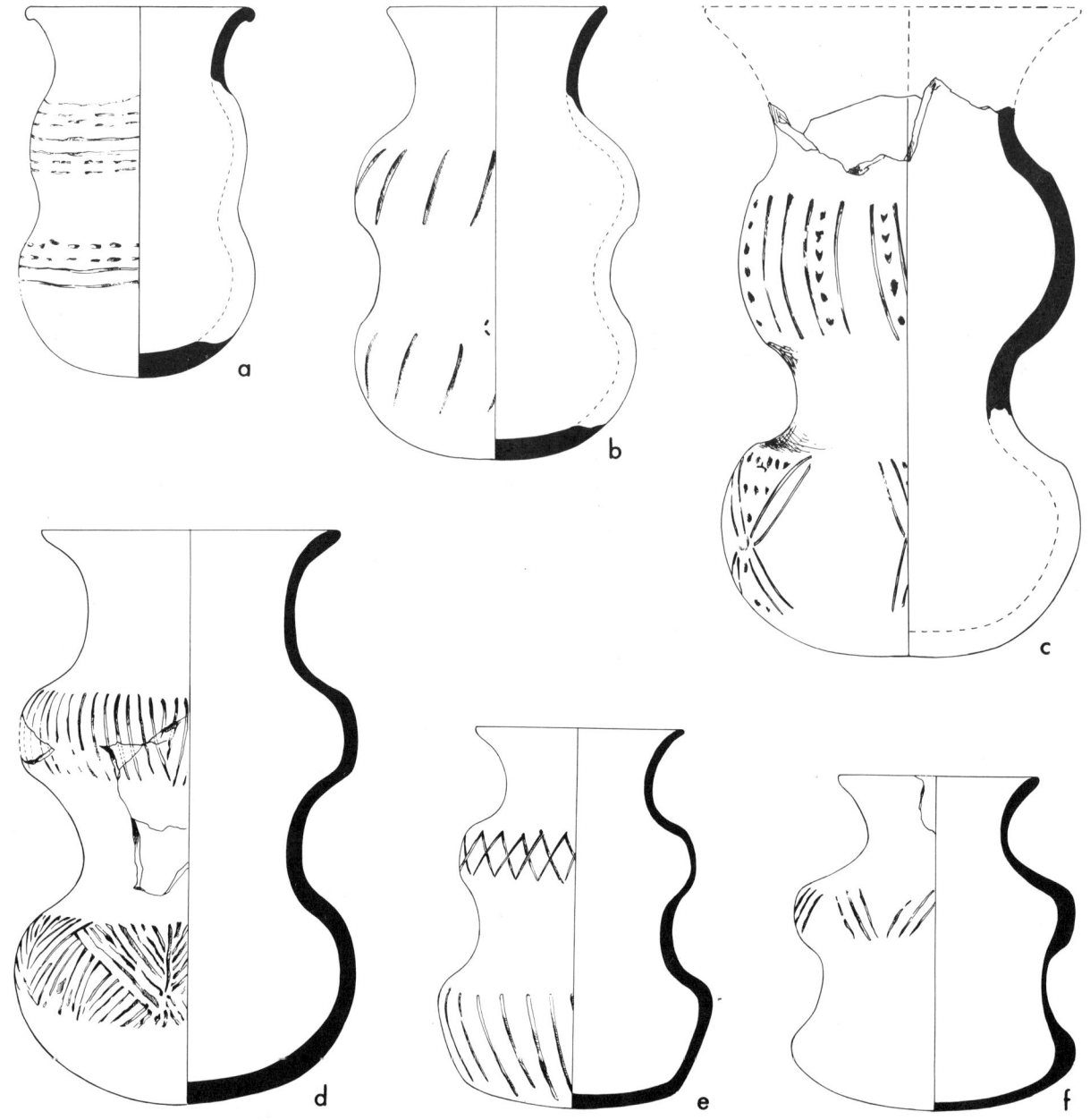

Fig. 20. Capacha monochrome. Decoration: incision and punctation;
various motifs. Bules. Scale 1/4. See pp. 54, 55; For *a, d, e,* p. 57.

a.   9617. Appendix IV-B-3. See also Fig. 7.   *d.* 10028. IV-E-10.
b.   8807. IV-E-6.   *e.* 8715 b. IV-E-6. See also p. 54.
c.   10029. IV-E-10. See also pp. 26, 58.   *f.* 8715 a. IV-E-6. See also Fig. 9 *i.*

Below, specimens will be mentioned individually, start-ing with two entire figures from La Cañada, recovered a couple of weeks after their sale to a dealer, and a third specimen, from Quintero, which was traced and repur-chased long after its sale.

1. Figurine (Fig. 34 *a–aaaa*). Female, seated, hands on belly. Body hollow, probably also the head; exagger-ated perforation at navel. Little detail except on head. Nose once prominent. Eyes are raised ovals, surrounding a circular pupil, which has a vertical slit. Mouth a raised oval, containing crude vertical incisions; general effect dis-tinctly gopherlike. Ears represented. At back of head, hair apparently clubbed or cut in a straight bob; same true of

specimens shown in Figures 34 *bbbb* and 34 *c* (rear of head not visible) (cf. Romano 1963: photo 22, for similar treatment on Tlatilco figurine). Headdress consists of two raised areas cut by vertical slits.

Rose paint in band across headdress and vertically on the ornaments of the latter; also about eyes and sides of nose, and above navel. Red stripe on upper arms, just above elbow, and one stripe on each stump that represents a leg. Paste coarse; firing poor, but specimen by no means unbaked.

2. Companion figurine (Fig. 34 *b–bbbb*). Likewise female; seated; stubby arms extended. Body hollow at least to neck, with perforation at navel. Nose broken. Eye

[68]

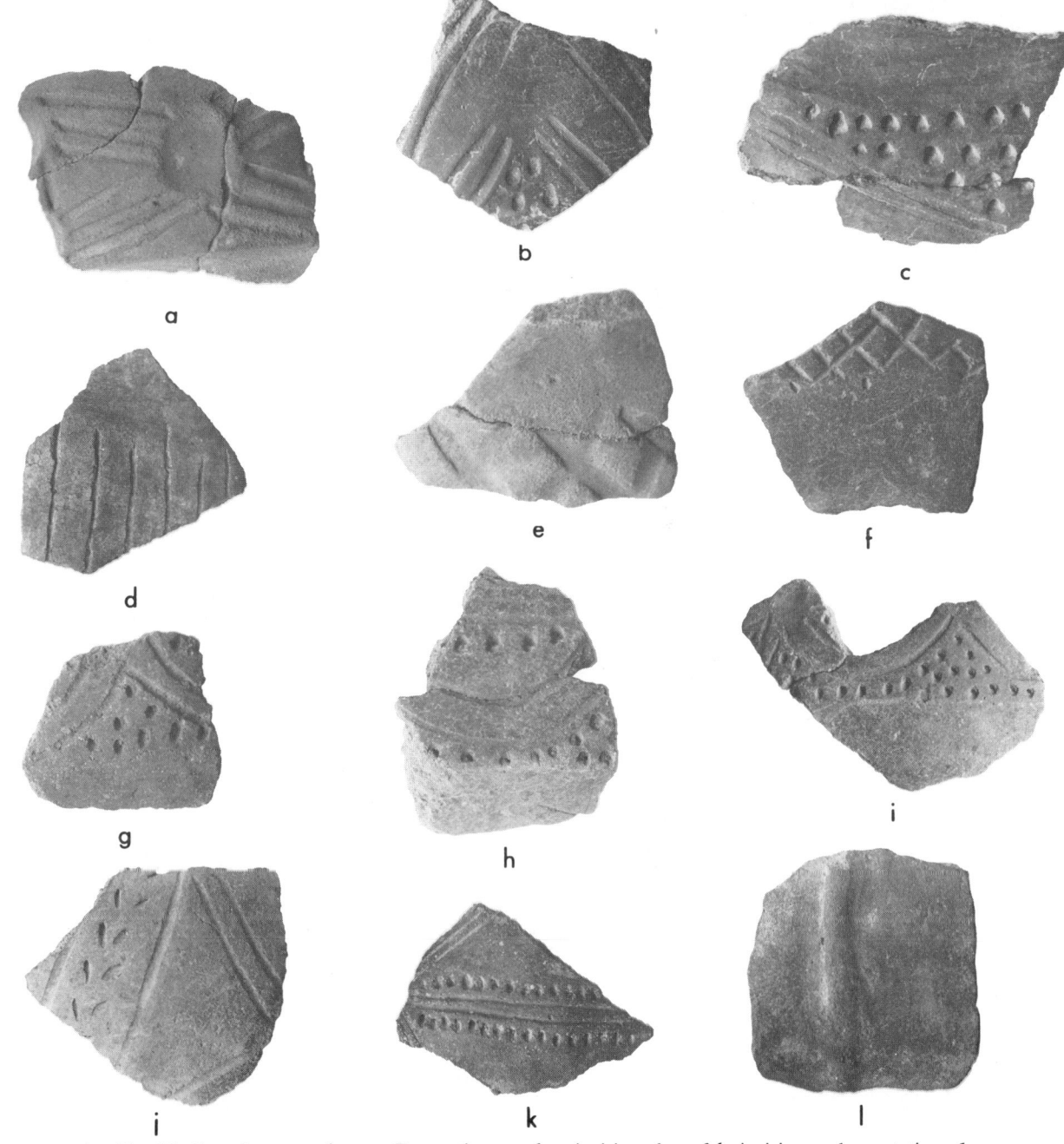

Fig. 21. Capacha monochrome. Decoration: *a, d, e,* incision; *b, c, f–k,* incision and punctation; *l,* modeling (raised rib). *b–d, j–l,* might be either bules or ollas; remainder, bules. Sherds; not in Appendix IV. Scale not uniform. For *a–c,* see also p. 26; *a, e,* p. 57.

a.	9960 a.	*c.*	8192 g.	*e.*	9960 h.	*g.*	9930 a.	*i.*	9930 b.	*k.*	9252 c.
b.	8207 b.	*d.*	9945 a.	*f.*	9913 m.	*h.*	9930 c.	*j.*	10217.	*l.*	10224 d.

and mouth treatment similar to that of preceding specimen, but form is more nearly almond shaped, with pointed corners. Mouth similarly gopherlike. Ears represented without detail. Headdress a simple raised area.

Rose paint across eyes and nose; vertical stripe extends upward and downward from each eye. Indeterminate smears of rose on shoulders, thorax, belly, and legs. Slight break in outline below belly and above subconical legs. Paste coarse; firing poor. Specimen has been broken at neck and mended.

3. A third figure (Fig. 34 *c–ccc*) likewise female; seated, with stubby arms extended. Differs little from pre-

ceding except in headdress; latter a raised horizontal band with opposed diagonal cuts, resulting in zigzag effect. No paint.

4. An additional specimen (Figs. 32 *b;* 33 *e*) fragmentary, but similar to the three preceding figurines. Sex not represented; hands on belly; fingers marked by three gouges. Head formed of superimposed rolls of clay; presumably body similarly constructed, but junctions obliterated. Body and head hollow; no perforation at navel. Eye and presumably mouth treatment similar to that of specimen 1. Head ornament uncertain; ill-defined splotches of rose paint.

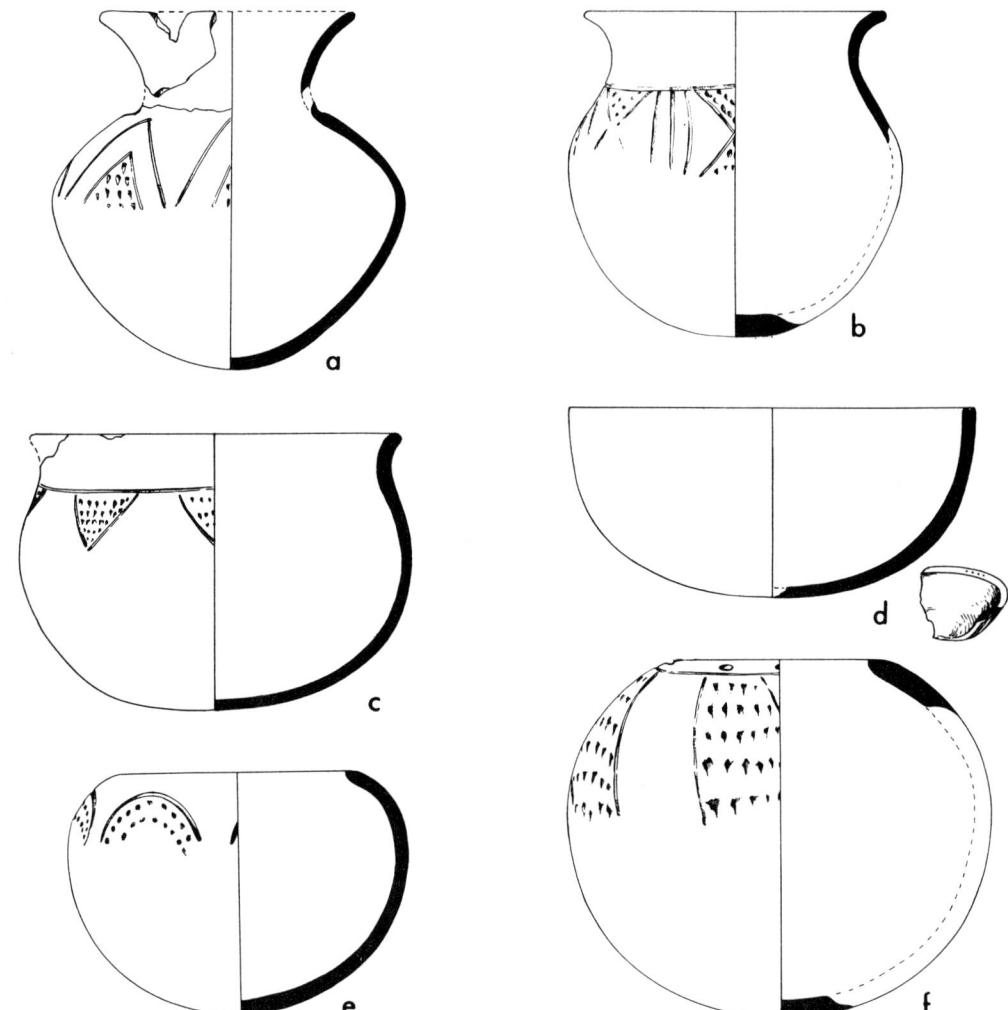

Fig. 22. Capacha monochrome. Decoration: incision and punctation; various motifs. *a,* Cántaro; *b, c,* ollas; *d–f,* bowls. Scale 1/4. For *a–c, e, f,* see also p. 58.

a. 9652. Appendix IV-B-6. See also p. 47; Fig. 10 *f.*
b. 9747. IV-A-10. See also p. 33; Figs. 8, 11 *j,* 23 *a.*
c. 10020. IV-E-10. See also p. 33.

d. 10021. IV-E-10. See also p. 24; Fig. 12 *d.*
e. 9716. IV-A-2. See also Figs. 5, 23 *c.*
f. 9749. IV-B-10. See also p. 24; Figs. 12 *h,* 23 *b.*

5.   A head fragment (Fig. 33 *b*) has a deep horizontal groove across forehead; above, a series of vertical, broad-line incisions; intervening raised surfaces painted rose.

6.   Another fragment (Fig. 33 *c*) may be part of the body belonging to the head just described, but the bits do not join. Seated, solid. A heavy roll of clay forms the central core of figure. Little detail; a single band of rose paint where the stump representing the leg joins the body. Additional small bits may belong to this, to preceding specimen, or to quite a different one.

7.   Another figure (Fig. 32 *c,* sketched from photograph of a piece in a private collection) has little detail but evidently relates to Capacha figures. Arms and legs no more than stubs joining an elongated body. Eyes slanting, with vertical slash; mouth curves upward on each side, its gopherlike aspect evident, especially in profile. Headdress represented by a series of vertical slashes. Notes say merely that object is of coarse clay, almost orange in color; no mention of red paint; no indication whether specimen solid or hollow, but probably the latter.

8.   A somewhat different style of figurine (Fig. 33 *a, aa*) is definitely unfired. Solid; standing, with stubby arms extended. A broad groove crosses the forehead; another groove at neck. Eyes the usual raised ovals, but without vertical slash. Nose quite prominent. Mouth treatment uncertain, but with a vaguely gopherlike aspect.

9.   The last of the Capacha figurines (Fig. 32 *a*) also is unfired. Apparently a hollow body, with flat, solid head and solid tips of extremities. Only the solid parts have been preserved. Horizontal groove across forehead is marked and seems to dip in middle, between the eyes. The one eye that remains is raised slightly and delineated by

Fig. 23. Capacha monochrome. Decoration: incision and punctation; various motifs. *a,* olla, with five design repeats; *b,* incurved bowl, six repeats, eight perforations; *c,* incurved bowl, six repeats. For *a–c,* see also p. 58.

a. 9747. Appendix IV-A-10. Rim diameter, 16.5–17.0 cm. See also Figs. 8, 11 *j,* 22 *b.*

b. 9749. IV-B-10. Interior rim diameter, 9 cm. See also Figs. 12 *h,* 22 *f.*

c. 9716. IV-A-2. Rim diameter, 12.0–12.5 cm. See also Figs. 5, 22 *e.*

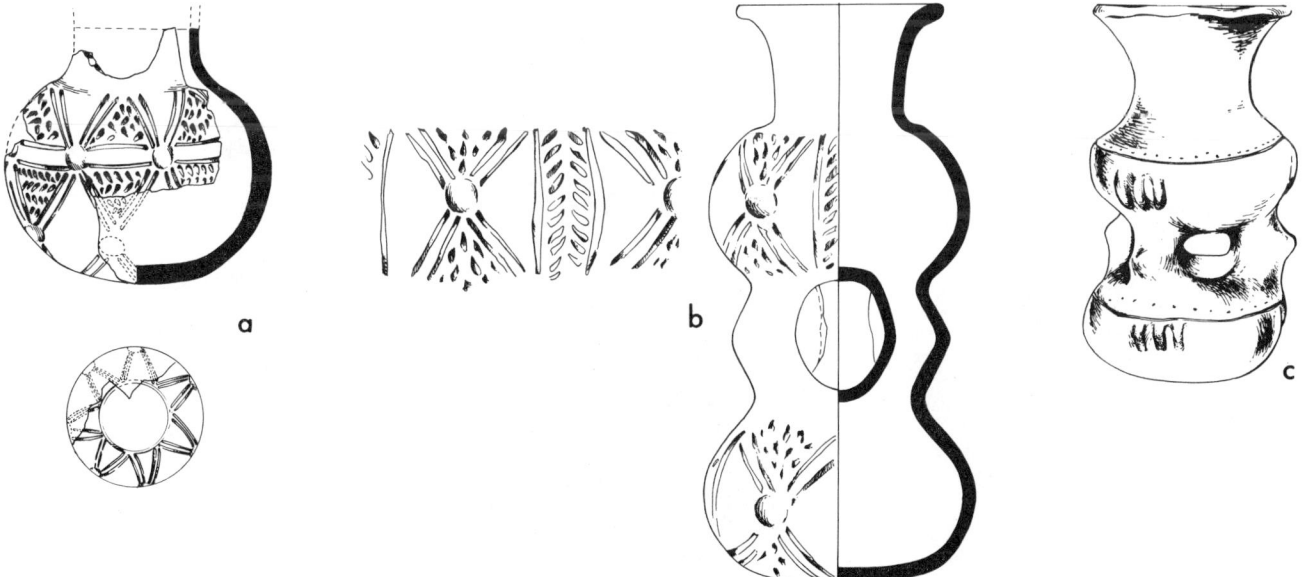

Fig. 24. Capacha monochrome. Decoration: *a, b,* incision and punctation, sunburst motif; *c,* incision and punctation, plus modeling. *a,* cántaro sherd with continuous sunburst forming lattice or network (p. 58); four repeats. *b, c,* stirrup vessels, three-tube variants (*b,* with sunburst and panels of opposed diagonal gashes, four repeats; *c,* with raised ribs and doughnut-shaped base). Scale 1/4.

a. 6321. Appendix IV-B, unnumbered site, Mesa del Salate (p. 94). See also pp. 26, 33, 35, 55, 58; Fig. 26 *a.*

b. 9970. IV-E-8. See also pp. 25, 26, 33, 49; Fig. 25 *c.*

c. Sketch from a photograph taken about 1940 of a specimen in private hands, Guadalajara, Jalisco; thought to be from Autlán, Jalisco. Not in Appendix IV. See also pp. 22, 25, 26, 33, 35, 36, 55, 56, 59.

[71]

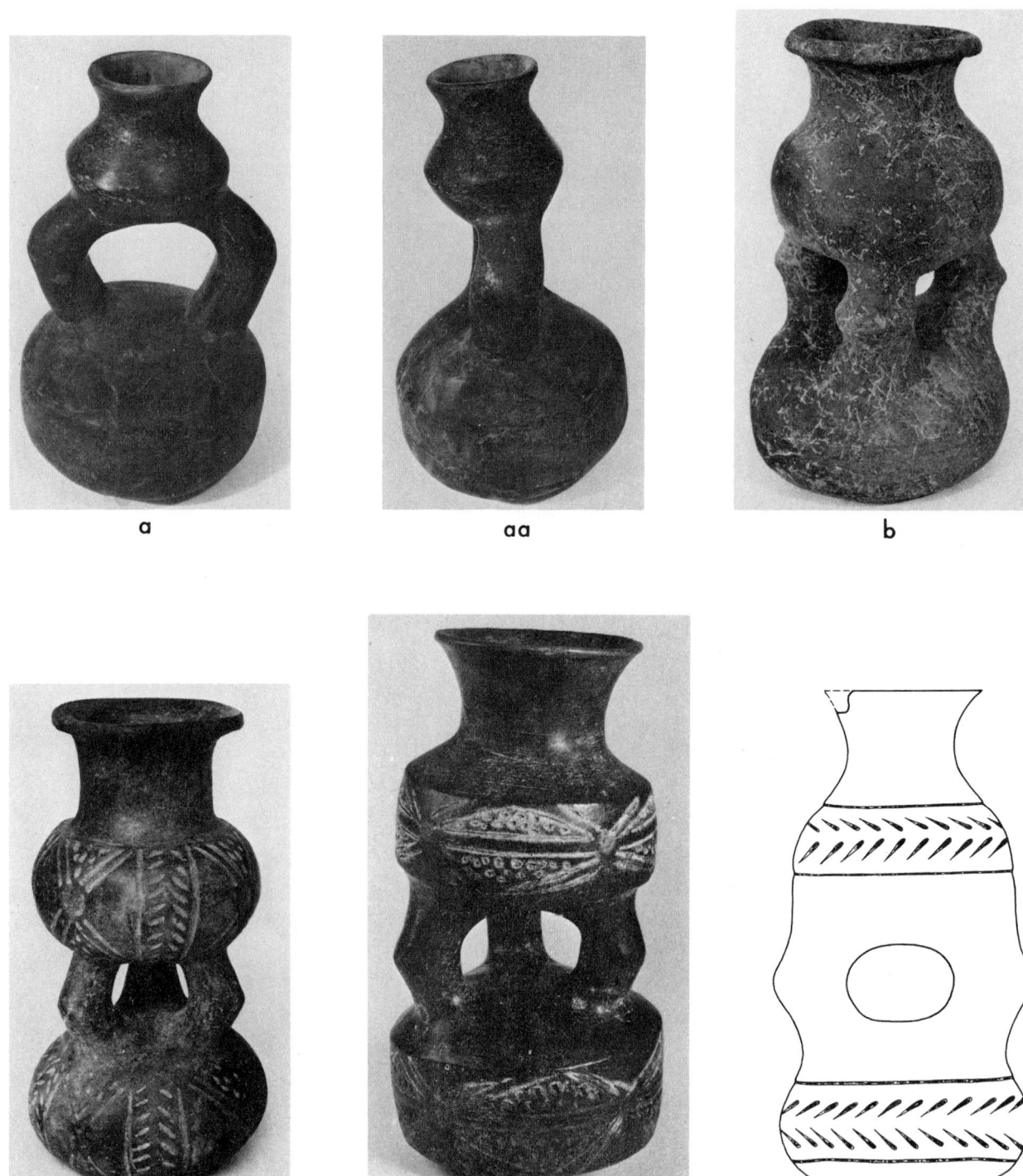

Fig. 25. Capacha monochrome. Vessel form and decoration: *a, aa, e,* stirrup pots, two tubes; *b–d,* variants with three tubes. Incision and punctation: *c,* sunburst alternating with panels of opposed gashes; *d,* sunburst variant, unusual in that punctation is to either side of navel, not above and below; *e,* two bands containing opposed gashes. For *a–d,* see also p. 55; *a–e,* pp. 25, 26; *b–d,* p. 33; *c–e,* p. 58.

a, aa. 9743. Appendix IV-A-10. Rim diameter, 7 cm. See also pp. 19, 51, 53; Figs. 8, 13 *d.*

b. 8714. IV-C-6. Rim diameter, 6.5–6.7 cm. See also Fig. 13 *e.*

c. 9970. IV-E-8. Rim diameter, 11.5 cm. See also pp. 26, 49; Figs. 2, 24 *b.*

d. Photograph courtesy of the Museum of the American Indian, Heye Foundation, no. 23/3034. Provenience unknown. Not in Appendix IV. Height, 28 cm. See also p. 26.

e. Specimen in the Museo Diego Rivera-Anahuacalli. Provenience unknown. Freehand sketch of vessel in glass case. Not in Appendix IV. Height guessed at 15-16 cm. See also pp. 35, 56.

Fig. 26. Capacha monochrome. Decoration: *a,* incision and punctation; *b–f,* modeling. *a,* continuous form of the sunburst resulting in a lattice or network effect (p. 58); *b,* finger grooving; *c,* raised ribs; *d,* ridged decoration; *e,* decorative protuberances at rim. *a–d,* cántaros; *e,* double bowl; *f,* effigy vessel. Not to scale; size indicated below for *b–f.*

a.   6321. Appendix IV-B, unnumbered site, Mesa del Salate (p. 94). See also pp. 26, 33, 35, 55, 58; Fig. 24 *a.*
b.   8708. IV-A-6. Rim diameter, 13.1–13.4 cm. See also pp. 18, 26, 43, 59.
c.   8713. IV-C-6. Rim diameter, 14.0–14.5 cm. See also p. 26; Figs. 10 *a,* 27 *a.*
d.   10138. IV-E-8. Rim diameter, 8.2–8.8 cm. See also pp. 26, 59.
e.   9632. IV-D-2. Maximum width of entire specimen, 24.6 cm. See pp. 40, 42, 59; Fig. 13 *b.*
f.   8706. IV-A-6. Rim diameter, 8.1–8.4 cm. See also pp. 18, 26, 49, 59; Fig. 13 *f.*

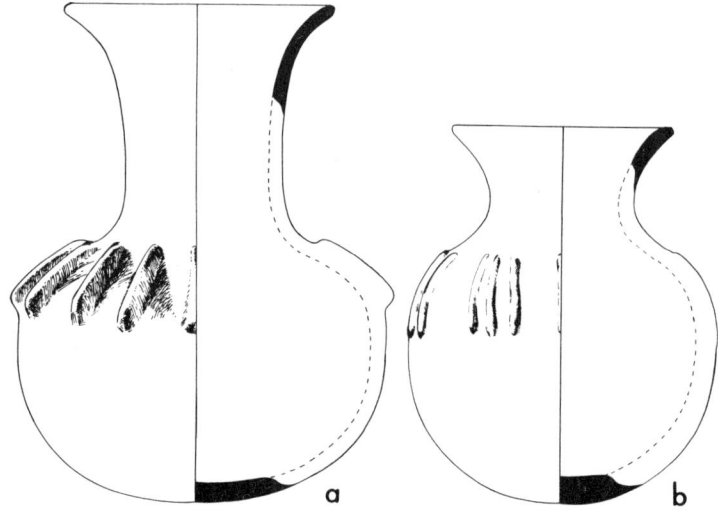

Fig. 27. Capacha monochrome. Decoration: modeling. Cántaros with raised ribs. Scale 1/4. See p. 55.
a. 8713. Appendix IV-C-6. Sixteen ridges, equispaced. See also pp. 26, 59; Figs. 10 *a*, 26 *c*.
b. 9687. IV-E-10. Five repeats of three ribs each. See also p. 26; Fig. 10 *b*.

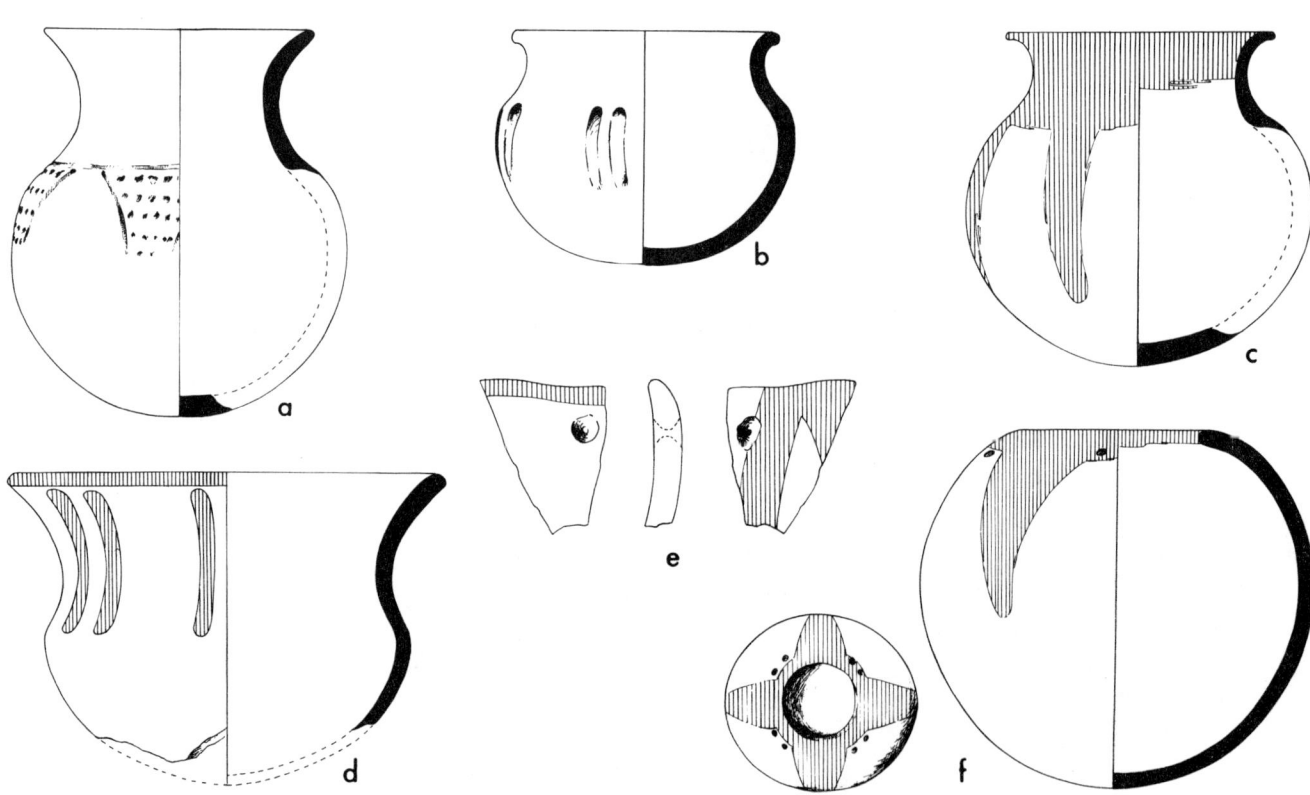

Fig. 28. Capacha monochrome: variants and dubiously classified. *a, c–f,* painted variants (p. 60); *b,* dubiously classified, perhaps not Capacha (p. 61). *a,* rose wash applied prior to incision and punctation; *b–f,* rose on natural brown; *b,* with finger grooving. *a–c,* ollas; *d–f,* bowls, various forms: *d,* flaring; *e,* pinched rim; *f,* incurved. Scale 1/4.

a. 9740. Appendix IV–B–10. Five repeats. See also p. 53; Fig. 29 *b*.
b. 9633 b. IV–D–2. Five repeats. See also pp. 26, 40, 43, 61; Fig. 29 *a*.
c. 9689. IV–E–10. Six repeats. See also p. 33; Fig. 30 *a*.

d. 9809. IV–B–3. Probably five repeats. See also pp. 33, 44; Fig. 6.
e. 9807 a. IV–B–2. See also p. 33; Fig. 31 *e*.
f. 9720. IV–A–2. See also pp. 24, 33, 42; Fig. 5.

Fig. 29. Capacha monochrome: variants and dubiously classified. *a,* aberrant, possibly not Capacha (described on pp. 61–62); *b–i,* variants with rose wash; *g,* partial slip. *a, b, f,* ollas; *h, i,* probably ollas; *c,* bule; *d, e,* probably bules; *g,* cántaro (rough interior). *c–h,* sherds, not in Appendix IV. Scale not uniform. For *a–i,* see also p. 26; *b–i,* p. 60.

a. 9633 b. Appendix IV–D–2. Five pairs of finger grooving. Rim diameter, 14.5 cm. See also pp. 40, 43, 61; Fig. 28 *b.*
b. 9740. IV–B–10. Five repeats; rose wash applied prior to incision and punctation. Rim diameter, 15 cm. See also Fig. 28 *a.*
c. 9960 j. See also p. 57. *d.* 9945 f. *e.* 10237. *f.* 9691 c. *g.* 9898 b.
h. 10259. See also p. 59.
i. 6322. Appendix IV–B, unnumbered site, Mesa del Salate (p. 94). See also p. 59.

broad-line incision; small punch in the middle instead of the common vertical cut. Ears represented. Faint vestiges of rose paint. Specimen must have been quite large.

In summary, Capacha figurines have little to recommend them aesthetically and some are not fired. Solid and hollow forms are known. Position may be standing or seated. Fingers and toes are rarely represented. Sex is not indicated invariably, but three entire specimens clearly are female (Fig. 34). There is some simple modeling to form a headdress and a bit of rose paint on the natural clay.

The figurines enumerated above have been examined by Profa. Rosa María Reyna, who feels that all may be considered Mesoamerican type K, specimen 8 above being "coarse or crude" K (K *burda*), while specimen 2, in

Fig. 30. Legend on top of page 77.

Fig. 30. Capacha monochrome painted variants and non-Capacha vessels for comparison.

a–e, Capacha monochrome, painted variants: *a,* red on brown, with red rim and throat, plus six vertical stripes; *b, bb, c, cc, d,* red on brown combined with incision; *b, bb, d,* plus punctation; *e,* red and black, zoned by incision. *f–i,* non-Capacha vessels not in Appendix IV: *f, h,* red on brown with incision; *g,* monochrome, reddish brown; *i,* trifid pot with red and orange overpaint on black (see p. 22). Not to scale; size indicated below. For *a–e,* see also p. 26; *a–d, g, h,* p. 33.

    a. 9689. Appendix IV–E–10. Rim diameter, 14.5–14.6 cm. See also p. 60; Fig. 28 *c.*

b, bb. 9626; *bb,* prior to restoration. IV–D–2. Rim diameter calculated at 24 cm. See also pp. 19, 26, 42, 60, Note 30 (p. 103).

c, cc. 9624. Restored. IV–D–2. Rim diameter calculated at 20.0–20.5 cm. See also pp. 19, 26, 42, 57, 59, 60.

    d. Private collection, Colima; provenience unknown. Rim diameter, 24.5–25.0 cm. Not in Appendix IV. See also p. 60.

    e. Museo Nacional de Antropología e Historia 2.3–459. Arroyo de San Antonio, Apulco, Tuxcacuesco, Jalisco (Fig. 1, no. 1). IV–B–1 (Kelly 1949: Pl. 14 *d*). Rim diameter, 18.0–18.5 cm. See also pp. 18, 31, 61.

    f. Museo Nacional de Antropología e Historia 2.5–5931. El Opeño, Michoacán. Maximum body diameter, 14.7 cm. See also p. 31.

    g. Museo Nacional de Antropología e Historia 1–1629. Tlatilco, Estado de México (p. 33). Rim diameter about 5 cm. See also p. 26.

    h. Museo Nacional de Antropología e Historia 1–1470. Tlatilco, Estado de México (p. 33). Rim diameter, 6.2 cm. See also p. 26.

    i. Museo Nacional de Antropología e Historia 2.1–166–58304. Possibly from Sinaloa (p. 22). Rim diameter, 3.8 cm. See also p. 22.

particular, has a slight suggestion of Mesoamerican type D-2. Resemblances to specimens from Morelos, in central Mexico, are mentioned elsewhere (Chapter 4: *"Tlatilco Style": Figurines*).

One figurine head (Fig. 31 *n*) not Capacha in style is illustrated because it suggests trade with coastal Guerrero; it comes from a Capacha cemetery but in mixed association.

Whistles

Two purchased ocarinas allegedly accompanied a Capacha burial. One (Fig. 35 *b*) is a bird effigy; the other (Fig. 35 *a*) is fragmentary, its original form uncertain.

Ornaments

Two unusual pottery beads (Fig. 35 *c*), funnel shaped, were allegedly found in Capacha association. Remarkably similar to one from Ticomán (Vaillant 1931: Pl. LXXXIV, second row, extreme left) and probably also to a specimen called an ear plug, from El Arbolillo (Vaillant 1935: Fig. 25, bottom, extreme right). Likewise in alleged Capacha association is a pottery pendant in the form of an animal (Fig. 35 *d*); its arched form suggests certain shell artifacts

from eastern Colima, where phases are still not disentangled.

Worked Sherds

Of five worked sherds (Fig. 31 *o–s*), four clearly are of Capacha monochrome. Four are roughly circular. One of these seems to have been the base of a bule, filed perhaps to serve as a plate or a cover; the other three are smaller. All are ill formed. The remaining fragment is quadrilateral, with rounded corners.

Unfired Clay Tablets

Two small, thin slabs of unfired marine clay (not illustrated; Appendix IV-A-2, 9715; IV-A-10, 9746) accompanied Capacha burials. One is roughly oval, measures 14 by 15 cm, and is approximately 7 mm thick. The other is smaller (6.5 by 8.4 cm) and about the same thickness; neatly formed, it is quadrilateral with rounded corners, and the edges taper.

Both specimens are of unknown utility. The Departamento de Prehistoria, Instituto Nacional de Antropología e Historia, identifies the material as fossilized limestone ($CaCo_3$) or "marine mud."

Fig. 31. Legend on top of page 79.

Fig. 31. Sherds and miscellaneous items from Capacha cemeteries. *a–j,* Capacha monochrome, atypical: *e–g,* painted variants; *e, f,* rose on brown; *g,* purple-black and rose, combined with incision; *h–j,* atypical in thinness and in circular punctation; *j,* better quality than the usual Capacha product, but apparently not a trade piece (p. 61). *k,* trade sherd, black, excised; source unknown. *l, m,* probably not Capacha; *l,* note lumpy, unsmoothed surface; *m,* comal fragment, underside (pp. 62–63). *n,* figurine head, trade, suggestive of coastal Guerrero style (p. 77). *o–s,* worked sherds.

Ordinarily, sherds are not included in Appendix IV; because of their special interest, *e–g, j* are exceptions. *k–n,* as non-Capacha artifacts in mixed association, do not appear in Appendix IV, and provenience is given below. *o–s,* worked sherds of Capacha monochrome, with the possible exception of *r.* Not to scale; size indicated below for *n* and *o–s.* For *a–d, h–j,* see also p. 59; *o–s,* p. 77.

a. 9945 c.
b. 9898 a. See also p. 59.
c. 9917 d. See also p. 57.
d. 9939 a.
e. 9807 a. Appendix IV–B–2. See also p. 60; Fig. 28 *e.*
f. 9807 b. Same as preceding. See also p. 60.
g. 9253. IV–B–7. See also pp. 49, 60.
h. 9918 a. Quintero, general digging. See also pp. 31, 59.
i. 9913 h. Same as preceding. See also p. 59.
j. 9961. IV–B–2.
k. 9815. IV–B–2, La Cañada, general digging. Association: mixed. See also pp. 29, 31, 43, 62.
l. 9876 a. La Parranda A, general digging. Association: mixed (Capacha to Chanal phases). See also p. 61.
m. 10346 a. Quintero, vicinity of Burials 8, 10. The sherd pictured was immediately above the bule (Fig. 17 *i*) that accompanied presumed Burial 8. See also pp. 51, 62, 63.
n. 9873. La Parranda A, general digging. Association: mixed (Capacha to Chanal phases). Maximum width, 4.2 cm. See also pp. 30, 44, 77.
o. 9926 b. IV–B–10.
p. 9926 a. IV–B–10.
q. 10239. IV–B–8.
r. 9926 c. IV–B–10. Height, 5.4 cm; *o–q, s,* to same scale.
s. 10226. IV–B–8.

Fig. 32. Capacha figurine and fragments. scarcely fired. *a,* slablike head fragment, with vestiges of rose paint, literally unbaked (p. 43); *b,* hollow fragment, faint smears of rose paint about eye, on headdress and chin; *c,* presumably hollow; not in Appendix IV. Association: *a,* securely Capacha; *b,* probably Capacha; *c,* none. Scale: *a,* ¼; *b, c,* ⅓.
a. 9723. Appendix IV–A–2. See also pp. 43, 70; Fig. 5.
b. 9625 b. IV–D–2. See also pp. 26, 33, 43, 69; Fig. 33 *e.*
c. Sketch from photograph of a specimen in a private collection, Colima. See also p. 70.

a   aa   b

c   d   e

Fig. 33. Capacha figurines and fragments. Unbaked to scantily fired. *a, aa,* solid, flat; *b,* headdress, with vestiges of rose paint; *c,* fragment showing construction based on a solid clay cylinder, position uncertain; cylinder may have been transverse at rump or vertical as shown; *d,* leg, better fired than other specimens here illustrated; *e,* fragment of a hollow specimen, showing construction of superimposed rolls of clay, visible especially in head area (exterior view, Fig. 32 *b*). Association: *a,* securely Capacha; *b–e,* probably Capacha. Not to scale; size indicated below. For *a, aa, b, c, e,* see also p. 43; *b, c,* p. 70; *e,* p. 69.

a, aa. 10294. Appendix IV–A–2. Height about 18 cm. See also p. 70.
b.   9625 c. IV–B–2. Maximum width, 7 cm. See also p. 27.
c.   9625 d. IV–B–2. Height as is, 12.2 cm.
d.   10247. IV–B–5. Maximum diameter as shown, 2.7 cm.
e.   9625 b. IV–D–2. Height, 11.8–12.0 cm. See also Fig. 32 *b.*

Fig. 34. Capacha figurines. Fairly well baked, but of kind considered unfired by moneros. Hollow. *a, b,* touches of rose paint; *c,* unpainted. Association: alleged, but quite certainly Capacha. Not to scale; size indicated below. For *a–ccc,* see also pp. 26–27, 33, 75; *a–bbbb,* p. 43; *b–ccc,* pp. 68–69.
a, aa, aaa, aaaa. 9649 a. Appendix IV–E–2. Eyes circled with rose paint, some of which extends over the chip on the nose, identifying the break as old damage. Height, 18.5 cm. See also p. 68.
b, bb, bbb, bbbb. 9649 b. IV–E–2. Height, 17.2 cm.
c, cc, ccc. 10323. IV–E–10. Height, 20.5 cm. See also pp. 53, 69.

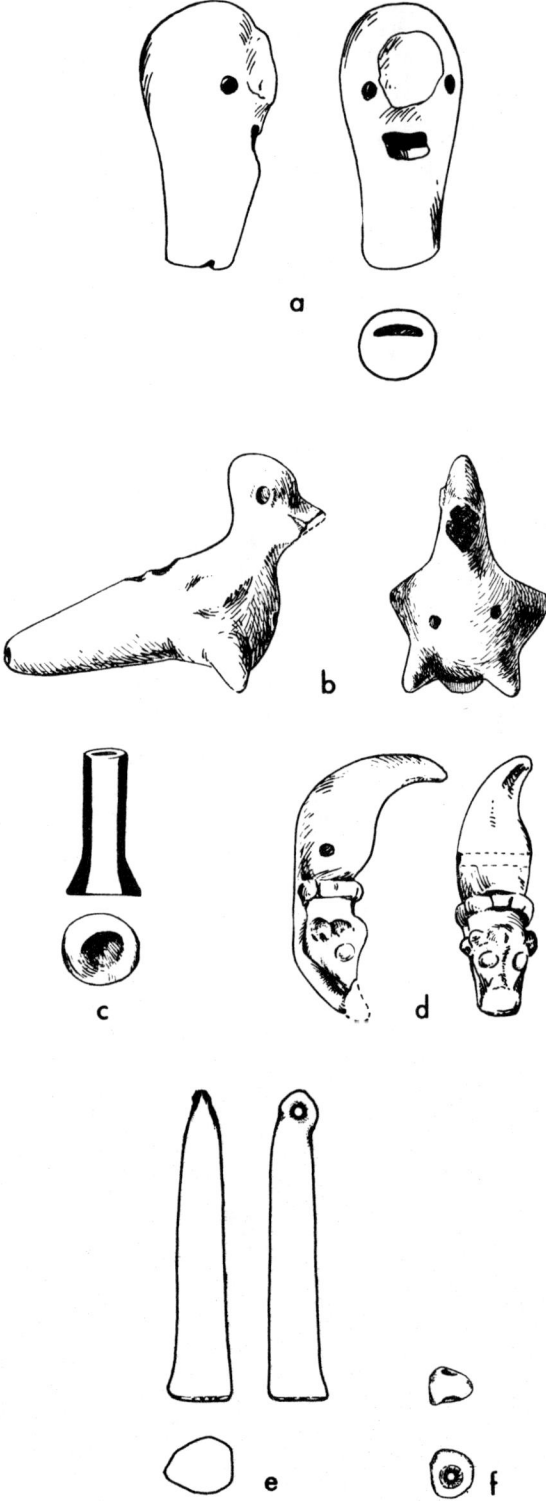

Fig. 35. Capacha products, miscellaneous. *a–d,* Ceramic; *e, f,* stone. *a,* Ocarina fragment; *b,* bird-effigy ocarina; *c,* funnel-shaped bead; *d,* effigy pendant; *e,* rock-crystal pendant; *f,* green-stone bead. In all cases, association probably Capacha. Scale: *c, e, f,* natural size; remainder, 1/2. For *a–d,* see also pp. 27, 77; *e, f,* p. 87.

a. 8680 b. Appendix IV–C–6.
b. 8680 a. IV–C–6.
c. 8719 a. IV–E–6.
d.   8709. IV–C–6.
e.   9994. IV–E–8.
f.   9993. IV–B–8. See p. 49.

Appendix III
STONE ARTIFACTS

All chipped stone is obsidian. Vesicular basalt—some fine grained, some coarse—was used for grinding or milling stones, manos, assorted receptacles, most "nutting" stones, and a number of oddments of dubious function. In addition, there are water-worn pebbles of various kinds of stone, probably of local provenience. However, one ax (Fig. 44 *a*), the poll of another (not illustrated; Appendix IV–B–8, 10221), and a large pebble used as some sort of polisher (Fig. 44 *c*) are all of *roca ígnea ultramáfica* (igneous rock high in magnesium and iron). It is possible that this material may have been collected outside the local area; at least, the few published data on Colima geology suggest that it is not plentiful in the zone. I am indebted to Ing. Adolphus Langenscheidt for guidance in the above statements and for specific identifications given below.

Capacha stone manufactures represent a certain anomaly. Chipped stone is limited to small chips or flakes of obsidian, a couple of scrapers, and, dubiously, a projectile point (Fig. 36 *f*). In contrast, ground-stone products are sophisticated and, in some cases, of skilled workmanship. Although most of the ground-stone artifacts were purchased, two were excavated under control and tend to confirm the alleged Capacha association of similar specimens.

OBSIDIAN

Most obsidian is black; several thin bits look light gray, and two specimens (one, however, of mixed association) are black shot with red. An informant in Chanchopa suggested one local obsidian source might be a deposit near Santa María de Miramar, on the Michoacán side of the Río Coahuayana and shown on one map at an elevation of about 400 m.[36] I had intended to visit the "mine," but my prospective guide left the area.

Some obsidian specimens from Capacha cemeteries are clearly non-Capacha in association; they are not included below or in Appendix IV. Associations of the remainder may be summarized thus:

1. Assuredly Capacha — 11
2. Probably Capacha — 5
3. Mixed, but chiefly Capacha — 9
4. Allegedly Capacha; purchased lot; reputed association probably correct because of comparatively early placement in Colima obsidian hydration series (Appendix IV 8720 a–k) — 11
5. Doubtfully Capacha; association mixed or scanty; not in Appendix IV — 25

   61

Most of the 61 specimens are small chips. Four such bits either have been retouched along the edge or have that appearance as the consequence of use (Fig. 36 *a–c*). There are two scrapers (Fig. 36 *d, e*), one of questionable Capacha association.

Prismatic blades are not a Capacha trait, nor, it appears, is surface retouching. However, there is one thin, well-made point (Fig 36 *f*) with slight lateral notches near the base; the precise form seems unusual in Mexico. Although association is not satisfactory (no. 5, above), conceivably this point might be a Capacha product. Another specimen, found in even more ambiguous association (likewise no. 5), is the fragment of a long, narrow retouched point of black and red obsidian (not illustrated), apparently about as thick as it is wide (1.2 cm).

GROUND STONE

Metates and Manos

Only one grinding stone, or metate (Fig. 37 *a*) comes in certain Capacha association. Eight others (Figs. 37 *c*, 38 *d*) are probably Capacha but association is not really secure.

With these considerable qualifications, it appears that Capacha grinding stones are oval or quadrilateral, with rounded corners; there is no sharp wall-base angle, and the exterior is well worked. The channel may be broad and shallow (Fig. 37 *b*), narrow and shallow (Fig. 37 *a*), or comparatively deep (Figs. 37 *c*; 38 *b, d;* plus two not shown, one of which has been used for red pigment). It appears that there was a preference for a long, narrow grinding slab, sometimes with marked trough. In addition, a nicely made miniature metate, quadrilateral, with sides bowed, has four nubbin feet, whose bases are cut at a peculiar angle (Figs. 39; 43 *d*); presumably a Capacha product, this specimen fits better with the stone receptacles mentioned below than with grinding stones.

The function of these metates is not very clear. One fragment mentioned above has been used for red paint. Of the others, only one (Fig. 38 *c*) is likely to have been serviceable for seeds, and the long, narrow, sometimes deeply channeled specimens may have had some special use. My housekeeper recalls having seen similarly proportioned milling stones used in Mazatepec, Morelos, for grinding cacao. Because the latter is costly, a reduced grinding surface is favored; otherwise much of the precious substance is wasted, embedded in the pores of a basalt metate of normal size and proportions.[37]

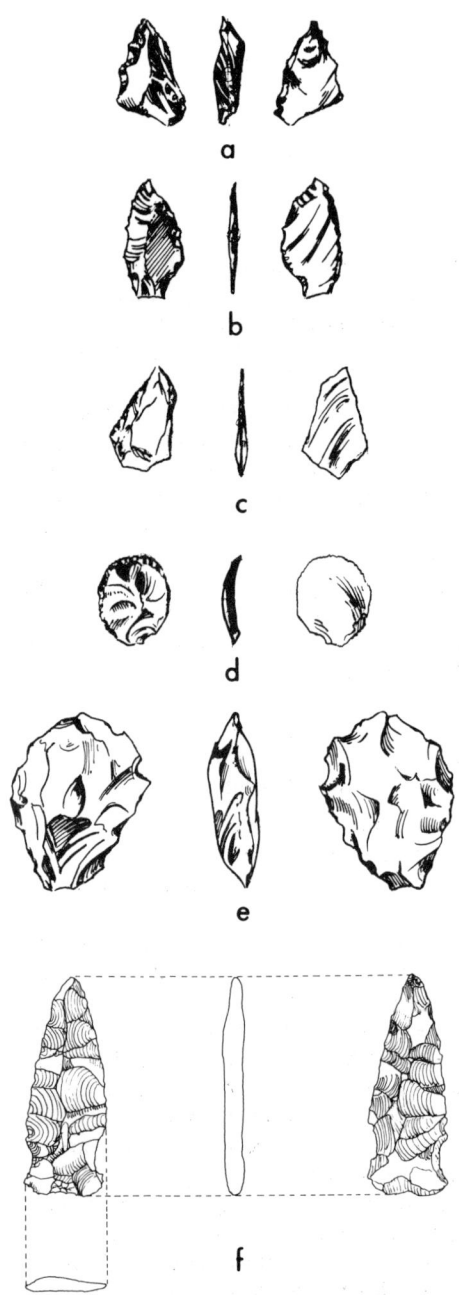

Fig. 36. Capacha obsidian artifacts. *a–c,* chips, edges retouched or nicked through use; *d, e,* scrapers; *f,* projectile point. Association: *a, b, e,* securely Capacha; *c, d,* probably Capacha; *f,* dubiously Capacha. Scale 1/2. For *a–f,* see also p. 83; *d–f,* p. 27.

a. 9744 *c.* Appendix IV–A–10.
b. 9744 *d.* IV–A–10.
c. 10236 *b.* IV–B–8.
d. 10236 *a.* IV–B–8.
e.   9745. IV–A–10. See also Fig. 8.
f.   9933. Not in Appendix IV. Association mixed, but possibly Capacha. From cemetery at La Capacha. See also p. 83.

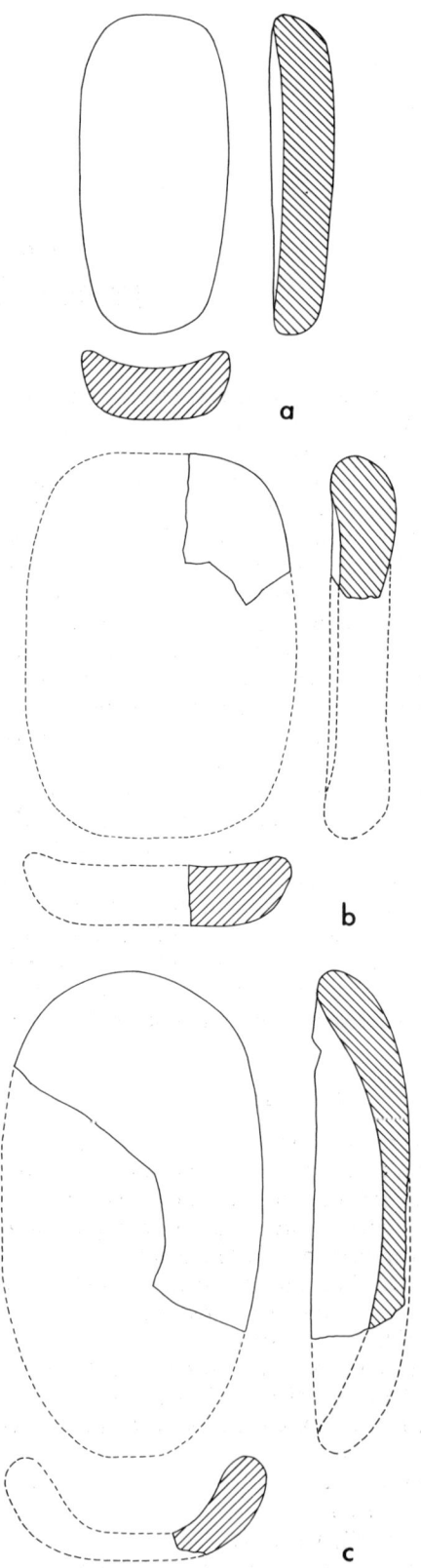

Fig. 37. Capacha metates. Association: *a,* securely Capacha; *b,* dubiously Capacha; *c,* probably Capacha. Scale 1/6. For *a–c,* see also p. 83.

a. 9748. Appendix IV–A–10. See also Fig 8
b. 9784. IV–B–10.
c. 9995. IV–B–8. See also pp. 50, 86.

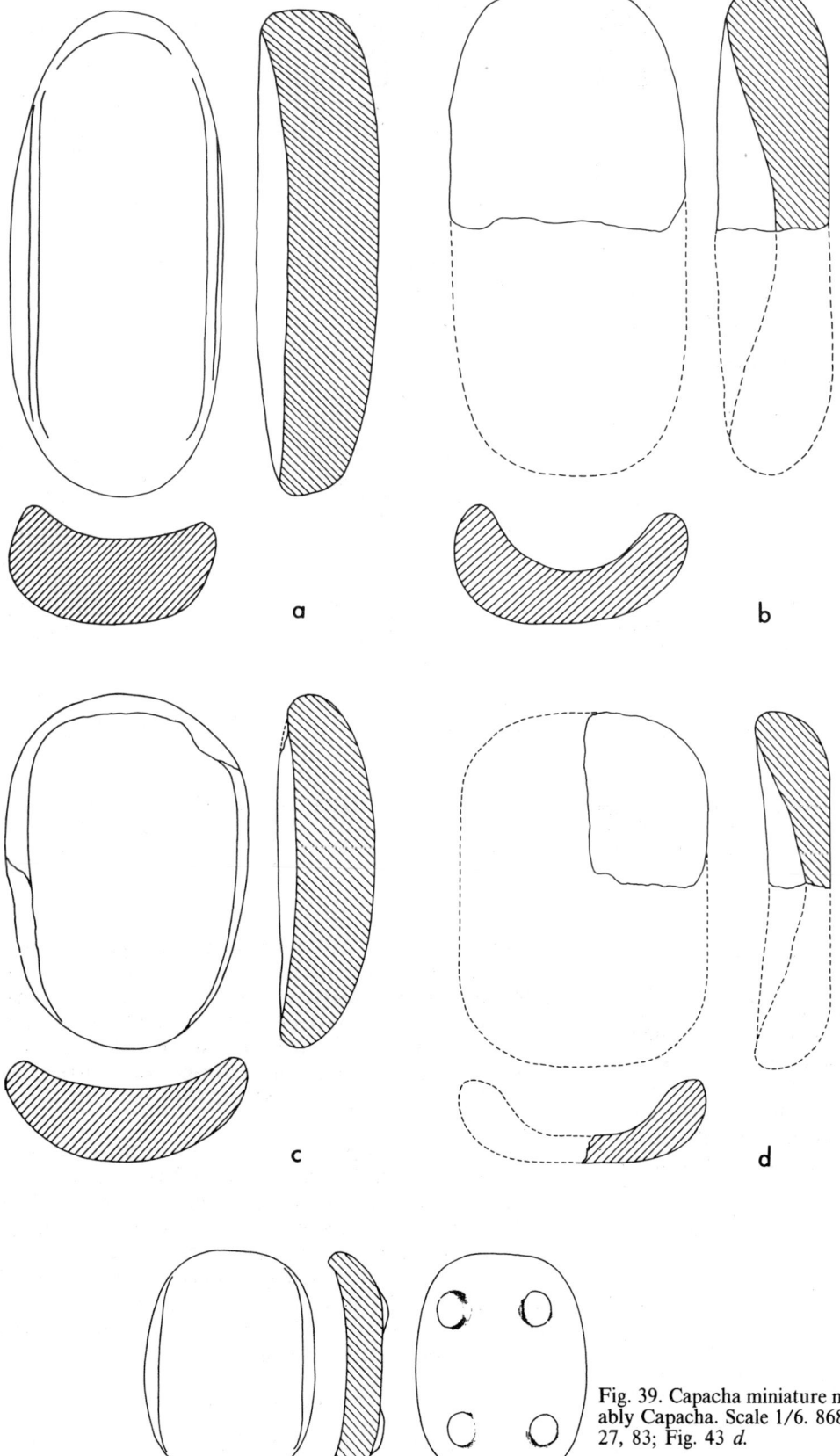

Fig. 38. Capacha metates. Note general resemblance between *a, c* and *b, d*. Association: *a–c,* dubiously Capacha; *d,* probably Capacha. Scale 1/6. See p. 27. For *b–d,* see also p. 83.

a. 10016. Appendix IV–E–10. See also p. 86.
b. 10314. IV–B–10.
c. 9692. IV–E–10. See also p. 83.
d. 9999 b. IV–B–8.

Fig. 39. Capacha miniature metate, tetrapod. Association: probably Capacha. Scale 1/6. 8682. Appendix IV–C–6. See also pp. 27, 83; Fig. 43 *d*.

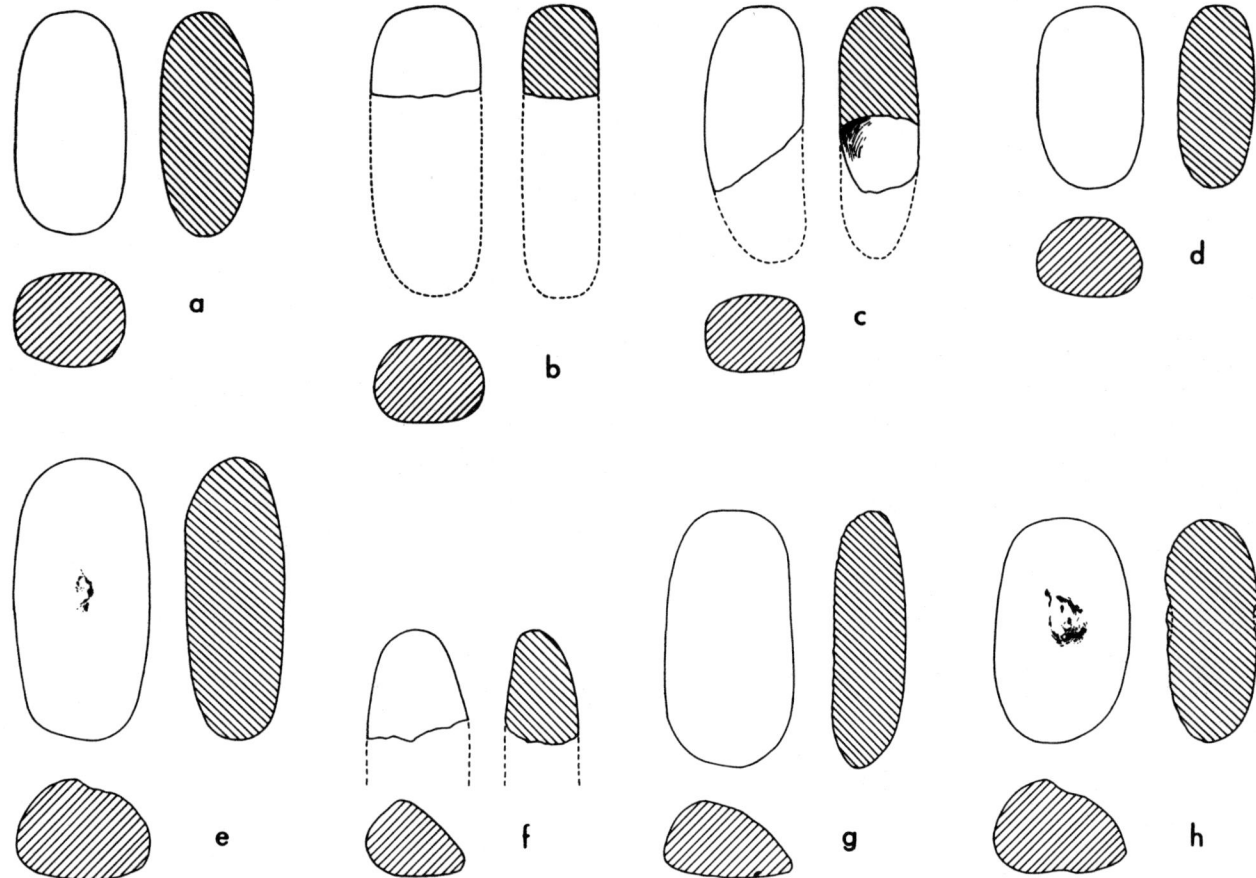

Fig. 40. Capacha manos. Association: *a–g,* probably Capacha; *h,* securely Capacha. Scale 1/6. See pp. 27, 86.

a. 9785. Appendix IV–B–10.	*c.* 9910 b. IV–B–10.	*e.* 10018. IV–E–10.	*g.* 9996 a. IV–B–8.
b. 9910 a. IV–B–10.	*d.* 10017. IV–E–10.	*f.* 9996 b. IV–B–8.	*h.* 10232. IV–B–8.

The few grinding stones attributable to Capacha have been viewed by Profa. Christine Niederberger. She sees a strong generic resemblance to specimens from Zohapilco, Estado de México—in the long, narrow, shallow grinding surface of some and in the relatively narrow, deeply troughed work surface of others. In her Zohapilco series (1976) these two forms continue side by side until Zacatenco times, when the patterns change. We both feel that such resemblances on an early time level must be significant and at least suggest specialized grinding operations presumably associated with food preparation. Unfortunately, neither of us can suggest a dichotomy of foodstuffs to account for the two basic metate forms.

The specific use of the grinding stone is particularly puzzling, because the nine manos that probably belong to the Capacha phase are unlikely companions for them. One mano, the fragment of another (Fig. 40 *f, g*), and a metate (Fig. 37 *c*) accompanied a burial, probably Capacha; the one mano is far too large for the grinding stone, and the fragment is small, size and proportions uncertain. Another hand stone (Fig. 40 *d*) was allegedly found with a metate (Fig. 38 *a*); the size is about right, but the mano has evidently been used to grind red paint, whereas the metate has not.

Of the manos or hand stones shown in Figure 40, only one (*h*) is assuredly Capacha. Two more (*f, g*) come from a burial without pottery, but probably of the Capacha phase. Other specimens (including one not illustrated) are even less securely allocated. It may be noted that all manos and metates come from two cemeteries: Terreno de Fidel Valladares (Fig. 1, no. *8*) and Quintero (Fig. 1, no. *10*).

The data scarcely permit generalization, but tentatively it may be said that the Capacha mano has a slightly convex grinding surface. The nongrinding face is often rather peaked (Fig. 40 *e–h*), resulting in a loaf-shaped or a nearly triangular section. Sides are not squared. One specimen (Fig. 40 *e*) has an irregular depression on the nongrinding face; another (Fig. 40 *h*) has a pronounced pit on both faces. Several manos show little evidence of wear and may even have been made for funerary use.

"Nutting" Stones

The so-called nutting stones (Fig. 41) are four—all biscuit shaped, with a shallow depression on one or both faces; one is of andesitic porphyry; the others are of vesicular basalt. The conventional "identification" may not be entirely arbitrary; today, on the Colima coast, similarly

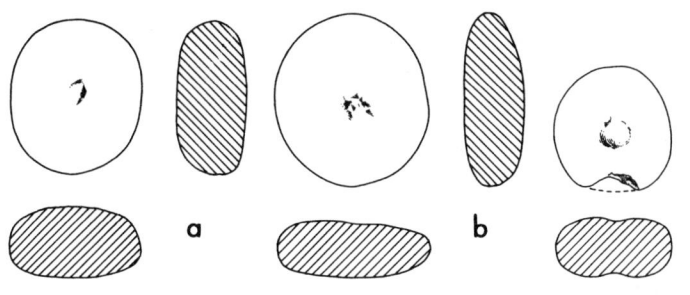

Fig. 41. Capacha "nutting" stones. Association: *a, b,* probably Capacha; *c,* securely Capacha. Scale 1/6. See also p. 86.
  a. 9909. Appendix IV–B–10.
  b. 9998. IV–B–8.
  c. 10235. IV–B–8.

pitted stones are used as anvils in breaking the nut of the *cayaco* palm. Archaeologically, Capacha association seems indicated; as is the case with manos and metates, all specimens come from two cemeteries (Fig. 1, nos. *8* and *10*).

One biscuit-shaped stone (not illustrated; Appendix IV–B–8, 10363) is generally similar but lacks the diagnostic pits. Of fine-grained vesicular basalt, it shows use on one face from some sort of rubbing or grinding operation.

Stone Receptacles

Several receptacles seem to be Capacha artifacts. Although only the first two of the five specimens listed below come from my testing, there seems no reason to doubt the Capacha attribution of the purchased pieces.

1. Oval plate with protuberance at one end, suggesting a "spoon" or "ladle"; well worked (Figs. 42 *a;* 43 *c, cc*).
2. Simple plate or bowl, roughly oval, but somewhat skewed; exterior quite well worked; side walls convex (Fig. 42 *b*).
3. "Bowl," roughly circular viewed from above; convex-sided. Shallow, well-worked depression does not show use as mortar (Fig. 42 *c*).
4. Stone dish, more or less circular; short tripod feet; well worked (Figs. 42 *d;* 43 *a, aa*).
5. Stone dish, oval; two cusped feet; well worked (Figs. 42 *e;* 43 *b, bb*).

Axes with Three-quarter Groove

Surprisingly, the ax with three-quarter groove is extremely rare in Colima, and the scanty data available suggest that it may be exclusively Capacha. This is puzzling because such artifacts are plentiful in the regional museum in nearby Ciudad Guzmán (Jalisco), there thought to be of local provenience. Otherwise, that museum has little reminiscent of Capacha, although it does have some unusual specimens of black pottery, including two vessels with cincture.

The only entire ax with three-quarter groove that I know from Colima (Fig. 44 *a*) was purchased in Capacha and reputedly had been found with a Capacha-phase burial. It is a nicely worked specimen of igneous rock that has been ground to a sharp edge; the latter shows little sign of use, except for a couple of large chips.

The polls of two specimens (not illustrated) indicate similar form; one is of andesite, the other, of igneous rock (Appendix IV–B–7, 8890, and B–8, 10221). Association is not secure.

Miscellaneous Manufactures

Several ground-stone artifacts are difficult to classify. One (Fig. 45 *a*) is a comparatively small, rough, palette-shaped object, apparently grooved for a raquettelike haft, or possibly for the smoothing of some sort of slender shaft. It will be remembered, however, that there is no indication of the use of the arrow during Capacha times. Another item of problematical use is the small corner from a quadrilateral object of diorite (not illustrated; Appendix IV–B–7, 8891); it may be from some sort of palette or small grinding stone.

Two small, palettelike slabs of limestone have rounded corners (Fig. 43 *f, g*); each has one face worn, as if it had been used for whetting; this face is slightly concave and in its central area there is fine, evidently deliberate pitting.

Still another object of uncertain use is the fragment of a notched stone of reddish basalt (Fig. 45 *b*), which suggests the specimens identified as figurines of the Ecuadorian Valdivia phase (Meggers, Evans, and Estrada 1965, Pl. 117 *r–t*).

A small, beautifully worked fragment of fine black non-vesicular basalt (Fig. 44 *b*) is of particular interest because of its resemblance to similar objects found elsewhere in Mexico in "early" context. Such artifacts sometimes are considered polishing stones used in pottery making, but it seems equally likely that they may have been amulets.

Two stone ornaments may be of Capacha workmanship; the association is probable but not certain. One is a slender pendant (Fig. 35 *e*) of rock crystal, pestle-shaped, smooth, but not very symmetrical; the perforation has been drilled from both sides, but there is considerable chipping about the aperture. The other is a small greenstone bead (of *crisoprasa*?), somewhat amorphous (Fig. 35 *f*); surprisingly, the bore has been drilled from one side only.

Water-worn pebbles—of various sizes and presumably of appealing appearance—were collected by Capacha people. Included are chalcedony, igneous rock, flint, rhyolite, and silicated(?) rock, the latter with veins of milky quartz and flint. Some have scarcely been altered; others have been somewhat shaped by rubbing; most are polished but show no signs of use. An exception, of igneous rock, has two faces worn smooth from service as a polisher (Fig. 44 *c*). It and a similar specimen (Appendix IV–B–10, 9912 a, b) are surface finds. Six small stones and pebbles (not illustrated; Appendix IV–C–6, 8683 a–f) allegedly accompanied a Capacha burial, and another (not illustrated; Appendix IV–B–8, 10228) comes from a test that is chiefly, but not exclusively, Capacha.

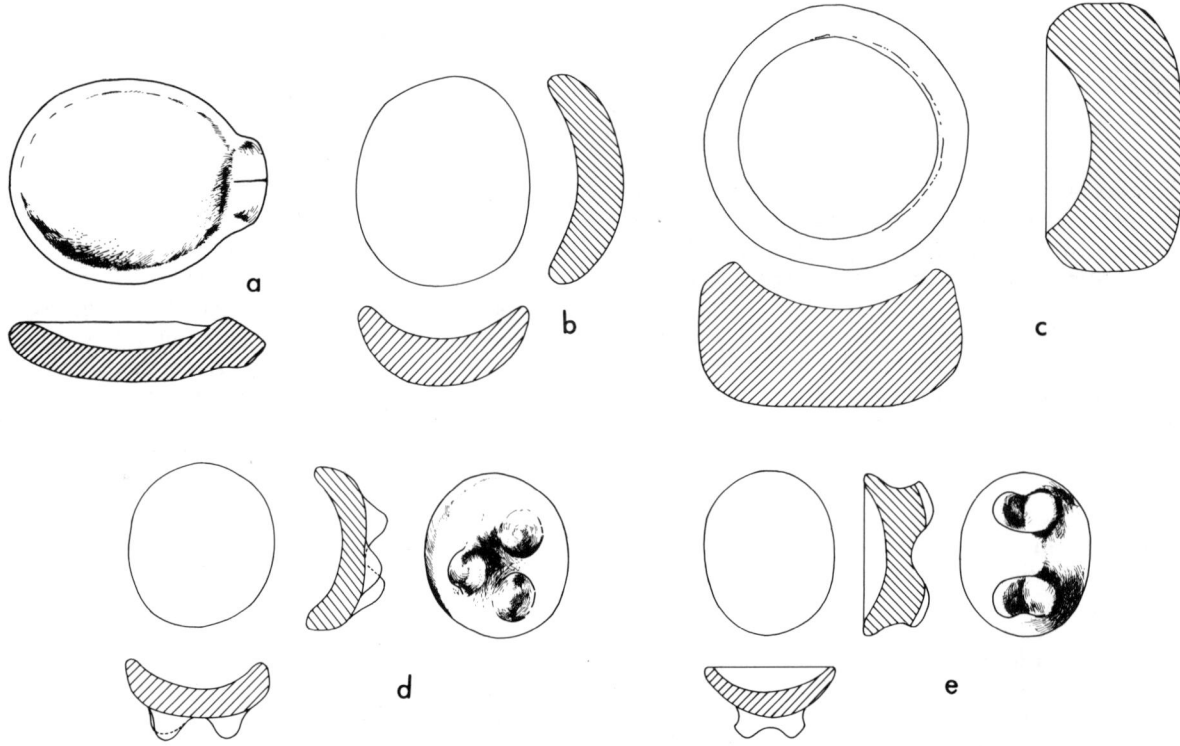

Fig. 42. Capacha stone receptacles. *a,* oval, one end worked to form a ladle or spoon; *d,* tripod; *e,* with two cusped feet. Association: *a, b,* presumably Capacha; *c–e,* probably Capacha. Scale: *a,* 1/4; *b–e,* 1/6. See p. 27. For *a–e,* see also p. 87; *a, b,* p. 47.

a.   9654. Appendix IV–B–6. See also Fig. 43 *c, cc.*
b.   9653. IV–B–6.
c.   9997. IV–E–8.
d.   8681. IV–C–6. See also p. 31; Fig. 43 *a, aa.*
e.   8686. IV–C–6. See also Fig. 43 *b, bb.*

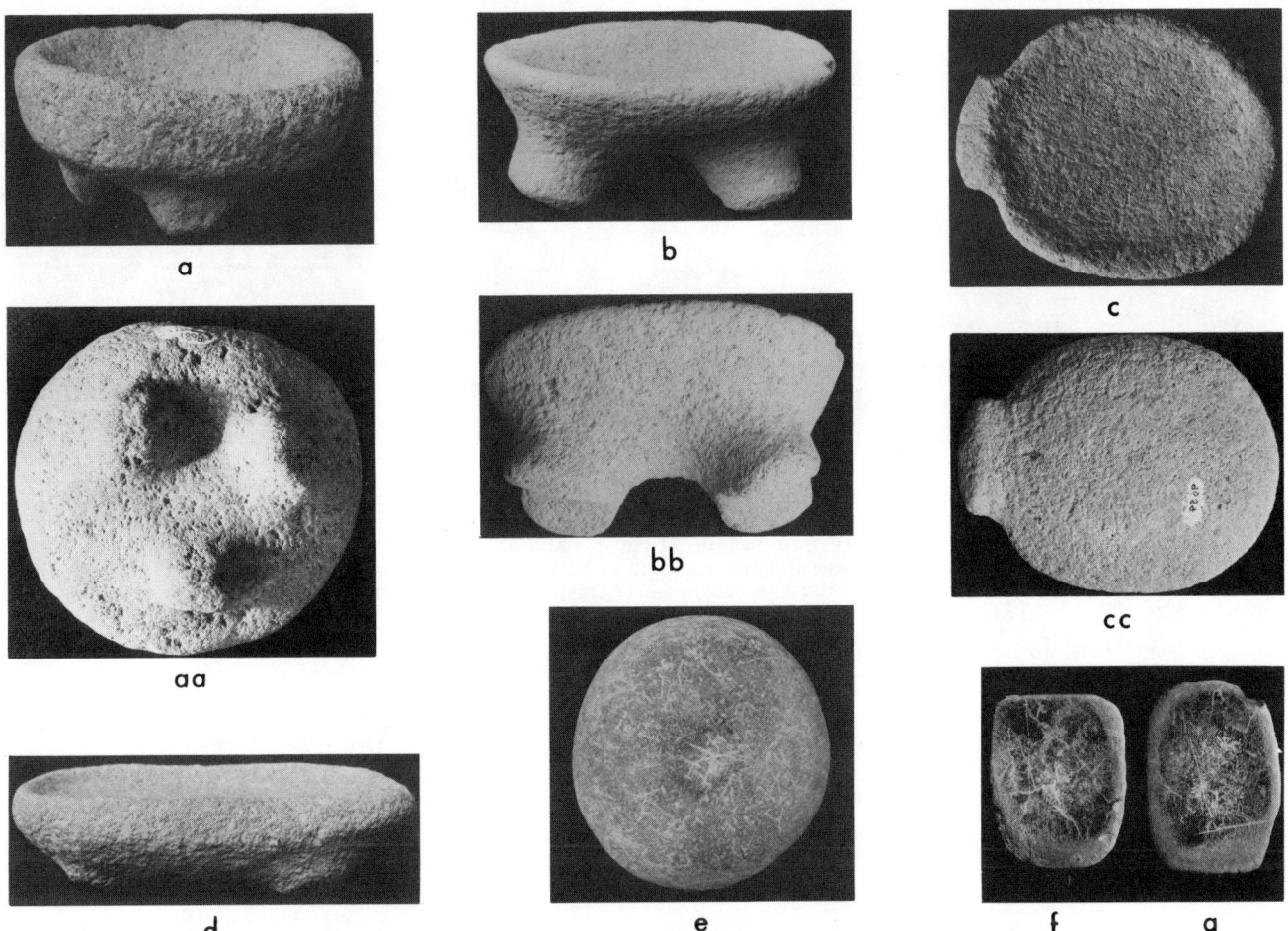

Fig. 43. Capacha stone artifacts. *a, b,* dishes (*a,* tripod; *b,* oval, with two cusped feet); *c, cc,* ladle or spoon; *d,* miniature metate, oval, with four nubbin feet; *e,* "nutting" stone; *f, g,* limestone "palettes," with fine, close pitting of surface. Association: *a, h, d–g,* probably Capacha; *c,* presumably Capacha. Not to scale; size indicated below. For *a–bb,* see pp. 27, 87.

    a, aa. 8681. Appendix IV–C–6. Maximum diameter, 12.8 cm. See also Fig. 42 *d.*
    b, bb. 8686. IV–C–6. Length, 12.9 cm. See also Fig. 42 *e.*
    c, cc. 9654. IV–B–6. Length, 14 cm. See also pp. 47, 87; Fig. 42 *a.*
    d. 8682. IV–C–6. Length, 20.7 cm. See also p. 83; Fig. 39.
    e. 8718. IV–E–6. Maximum diameter, 8.8 cm.
    f, g. 9973 a, b. IV–E–8. Length of *g,* 10.2 cm. See also p. 87.

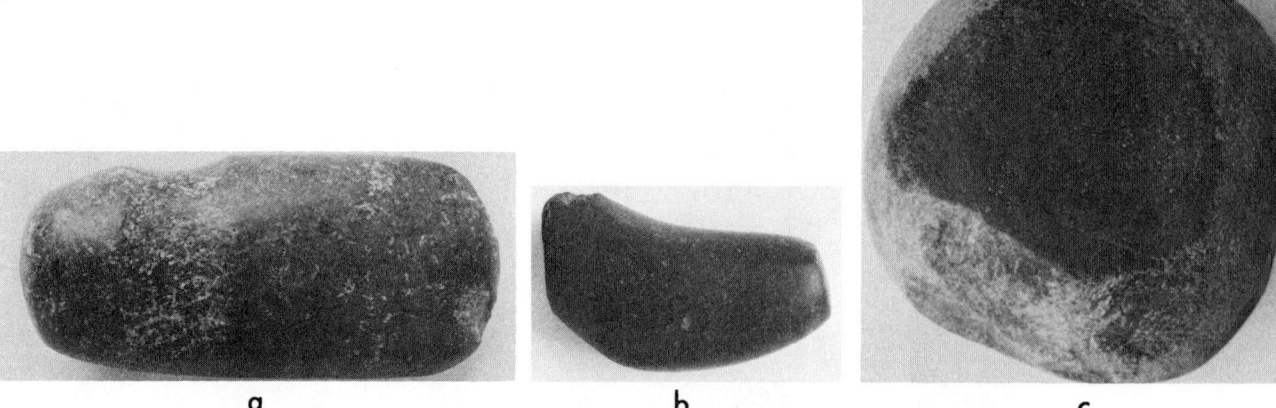

a   b   c

Fig. 44. Capacha stone artifacts. *a,* ax; *b,* amulet(?); *c,* polisher. Association: *a,* presumably Capacha; *b, c,* possibly Capacha. Not to scale; size indicated below.

a.   8687. Appendix IV–C–6. Length, 11 cm. See also pp. 27, 83, 87.
b.   10344. IV–B–10. One tip missing; length as is, 4.5 cm. See also pp. 27, 34, 87.
c.   9912a. IV–B–10. Diameter approximately 6 cm; thickness, 4.8–5.2 cm. See also pp. 83, 87.

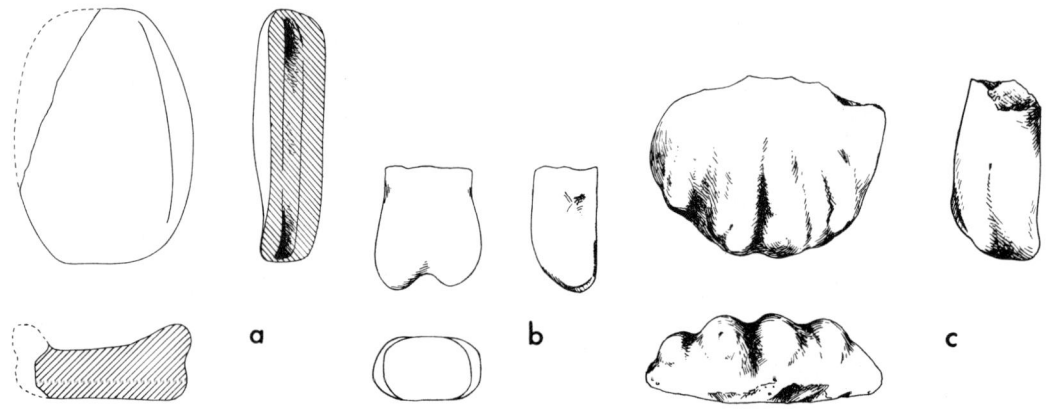

a   b   c

Fig. 45. Capacha stone artifacts. *a,* palettelike object, grooved for hafting; *b,* notched stone; *c,* animal foot(?). Association: *a, c,* dubiously Capacha; *b,* probably Capacha. Scale 1/4.

a.   9908. Appendix IV–B–10. See also p. 87.
b.   10222. IV–B–8. See also p. 87.
c.   9911. IV–B–10.

Appendix IV
PROVENIENCE AND ASSOCIATION

In this appendix, Capacha materials are grouped according to provenience and association. Categories run from A through E, more or less in descending order of reliability. A refers to grave lots; B, to cemetery association; C, to alleged grave lots; D, to alleged monero-pit association; E, to reputed cemetery provenience.

Following the lettered categories comes a number that identifies the specific Capacha cemetery; the numbers are the same as those used in Figure 1 and Appendix I. Thus A-2 refers to grave lots from La Cañada; A-3, to similar lots from La Parranda A. For want of information, no number is assigned two localities which appear, respectively, at the end of the B and E entries.

After the site designation comes the field-catalog number. These numbers appear in the order in which specimens were cataloged, hence they are not invariably in strict numerical sequence. For example, under A-3, the one specimen which accompanied Burial 5 is 9612, while 9609–9611 and 9613–9616 apply to Burial 4. There are few such instances, and the groups are quite small, hence there should be no difficulty in locating a given number. The legends accompanying figures 5 through 45 give precise references to this appendix, so that provenience and association can be ascertained rapidly and painlessly.

Except in cases of special interest, sherds are not included. In contrast, because of its potential importance in dating and in revealing the sources of a primary material, obsidian is fully listed, except for examples of dubious or inconclusive association. With few exceptions, obviously non-Capacha material does not appear in this appendix, even though it is attributed to a Capacha cemetery.

The entire Capacha collection has been delivered to the Museo Nacional de Antropología e Historia. Catalog numbers applied by the Museo can be ascertained by checking the field numbers herein against the catalog of the Museo.

A. GRAVE LOTS

Obviously, grave-lot association is the most reliable although, in some instances, bones of several individuals are involved. Entries under A refer to my excavations and observations.

A-2. La Cañada

Group Burial 2 (two or more individuals; Fig. 4).
9712. Capacha monochrome, bule, undecorated (Figs. 4, no. *6*; 9 *h*).

9713. Capacha monochrome, bule, decorated (Figs. 4, no. *7*; 15 *a*; 17 *b*).
9714. Capacha monochrome, bule neck; may have accompanied an earlier burial in the same subsoil pit (Fig. 4, no. *8*).
9715. Thin slab of fossilized limestone (CaCO₃); beneath 9714 (Fig. 4, no. *9*) (see Appendix II: *Unfired Clay Tablets*).

Burial 3 (more than one individual, or major dislocation owing to a subsequent interment; Fig. 5, no. *2*).
9716. Capacha monochrome, incurved bowl, decorated (Figs. 5, no. *4*; 22 *e*; 23 *c*).
9718. Capacha monochrome, olla, undecorated (Figs. 5, no. *3*; 11 *g*).
9723. Capacha figurine fragment, semifired. Evidently large, but only a head fragment remains, plus solid tips of extremities (barely visible in Fig. 5, no. *6*; Fig. 32 *a*). Probably accompanied Burial 3 and later was covered by Burial 5 (see latter, below); found immediately beneath Capacha bule 9722.

Burial 4 (one individual; Fig. 5, no. *7*).
9719. Capacha monochrome, bule, decorated (Figs. 5, no. *8*; 17 *e*).
9720. Capacha monochrome, painted variant; incurved bowl (Figs. 5, no. *9*; 28 *f*).

Burial 5 (secondary burial within a bule).
9722. Capacha monochrome, bule, decorated; used as burial urn (Fig. 5, no. *5*).
10293 a, b. Fragmentary remains of two adults, within bule 9722 (Appendix V, no. 10293 a, b).
10294. Capacha figurine fragment, unfired (Fig. 33 *a, aa*); found within bule 9722.

A-3. La Parranda (Site A)

Burial 2 (not illustrated).
9605. Capacha monochrome, bule, decorated (Fig. 19 *a*).
Burial 3 (Fig. 6, no. *5 a, b*).
9607. Capacha monochrome, bule, decorated (Figs. 6, no. *7*; 16 *a*).
9608. Capacha monochrome, olla, undecorated (Figs. 6, no. *6*; 11 *i*).
Burial 4 (Fig. 7, no. *1*).
9609. Capacha monochrome, bule, decorated (Figs. 7, no. *3*; 15 *d*; 17 *d*).

9610. Capacha monochrome, bule, decorated (Fig. 7, no. *4*).

9611. Capacha monochrome, bule, decorated (Fig. 7, no. *5*).

9613. Capacha monochrome, bule, decorated (Fig. 7, no. *2*).

9614. Capacha monochrome, bule, undecorated (Fig. 7, no. *7*).

9615. Capacha monochrome, bule, decorated (Fig. 7, no. *8*).

9616. Capacha monochrome, bule, decorated (Fig. 7, no. *6*).

Burial 5 (Fig. 7, no. *10, 10 a*).

9612. Capacha monochrome, decorated; large sherd (Fig. 7, no. *11*).

A-5. Parcela de Luis Salazar

Burial 1 (not illustrated).

10242. Capacha monochrome, cauldron, undecorated (Fig. 12 *c*).

10243. Presumably Capacha monochrome, miniature olla, undecorated, poor quality (Fig. 14 *d*).

A-6. La Capacha

At La Capacha, I witnessed a monero remove all but one of the specimens listed below and can vouch for grave association. The few isolated artifacts I found in general digging at the same site are listed below, under B-6.

Burial unnumbered, monero excavation

8676 a, b. Obsidian chips. Association firmly Capacha.

8677. Capacha monochrome, olla, undecorated (Fig. 11 *b*).

8678. Capacha monochrome, bowl, undecorated, pinched rim; semirestorable (not illustrated).

8706. Capacha monochrome, bird-effigy pot (Figs. 13 *f*; 26 *f*).

8707. Capacha monochrome, bule, decorated (Figs. 9 *d*; 19 *c*).

8708. Capacha monochrome, cántaro with finger grooving (Fig. 26 *b*).

A-7. El Barrigón

Burial, unnumbered.

8632. Fragmentary cranium (Appendix V, no. 8632; Figs. 46, 47).

8633. Capacha monochrome, cross-hatch decoration (Fig. 19 *b*).

8634. Obsidian chip, accompanying 8633. Association firmly Capacha.

A-10. Quintero

Burial 4 (Fig. 8, no. *2*).

9743. Capacha monochrome, stirrup vessel (Figs. 8, no. *3*; 13 *d*; 25 *a, aa*); beneath skull of Burial 2; probably associated with Burial 4.

9744 a–f. Obsidian chips, immediately beneath 9743 (9744 c, Fig. 36 *a*; 9744 d, Fig. 36 *b*). Association firmly Capacha.

9745. Obsidian scraper (Fig. 36 *e*); at same level as 9744, but 15 cm to the east. Same association.

9746. Oval slab, 7 mm thick, of material identified by the Departamento de Prehistoria as: "Caliza fosilífera microcristalina. Los fósiles son foraminíferos del grupo de las globigerinas; corresponden al Terciario. Sedimento marino; presenta también lentículas de óxidos de hierro" (not illustrated; see Appendix II: *Unfired Clay Tablets*). Burial 5 (Fig. 8, no. *4*).

9747. Capacha monochrome, small olla, decorated (Figs. 8, no. *5*; 11 *j*; 22 *b*; 23 *a*).

9748. Stone grinding slab (metate) (Figs. 8, no. *6*; 37 *a*).

9752. Incomplete skull and incomplete mandible of child burial (Appendix V, no. 9752).

Burial 7

10300. Mandible and skull fragments (Appendix V, no. 10300).

10301. Capacha monochrome, bule, undecorated; "killed" (not illustrated).

Burial 8

10302. Capacha monochrome, bule, unusually wide grooving, without punctation (Fig. 17 *i*).

Burial 9

10303. Capacha monochrome, bule, undecorated; widely flaring rim (Fig. 9 *e*).

Burial 10

10304. Capacha monochrome, basin, undecorated (Fig. 12 *a*).

10305. Skeletal remains, incomplete female interment (Appendix V, 10305).

B. CEMETERY ASSOCIATION

Several kinds of cemetery association are included in this lot. On the one hand, reference may be to material from the surface or from general digging at a Capacha cemetery, but without specific grave association. Inasmuch as such burial grounds were used successively, one Capacha interment often disturbed earlier ones, so that the occurrence of isolated specimens is quite common.

On the other hand, reference may be to monero discards collected on the surface of a rifled Capacha cemetery. Such material can be attributed safely to that burial ground, for moneros do not transport broken specimens from one place to another. A good many restorable and semirestorable vessels have precisely such provenience.

In addition, it must be added that most Capacha cemeteries are not exclusively of the phase of that name. Although provenience is secure, cultural association may be mixed.

Allocation of material to the present section is based on my excavations and observations.

B-1. Arroyo de San Antonio

The material entered below was reported some 25 years ago. At that time, it was unclassified but now can be related confidently to the Capacha phase. The sherds I collected on the surface; the restorable and near-restorable specimens were purchased locally.

Kelly 1949: Fig. 59 *b*. Apparently Capacha monochrome; sherds; ornament unique, but channel at base of neck quite typical.

Kelly 1949: Fig. 59 *c*. Capacha monochrome, bule, decorated; rim missing.

Kelly 1949: Pl. 14 *d*. Capacha monochrome, painted variant; black and red, with white-filled incision (Fig. 30 *e; Museo Nacional de Antropología no. 2.3-459*).

B-2. La Cañada

9625 c, d. Capacha figurine fragments (Figs. 33 *b, c*), little fired; not restorable. Surface, monero discards.

9711. Capacha monochrome, small olla, undecorated (Fig. 11 *h*). General digging.

9797. Capacha monochrome, bule, undecorated (Fig. 9 *g*). Surface, monero discard.

9798. Capacha monochrome, bule, decorated (Figs. 16 *c; 18 e*). Surface, monero discard.

9799. Capacha monochrome, bule, decorated (not illustrated). Surface, monero discard.

9800. Capacha monochrome, bule, decorated (not illustrated). Surface, monero discard.

9801. Capacha monochrome, bule, decorated (not illustrated). Surface, monero discard.

9802. Capacha monochrome, bule, decorated (not illustrated); semirestorable. Surface, monero discard.

9803. Capacha monochrome, bule sherd, decorated (Figs. 16 *b; 18 a*). Surface, monero discard.

9804. Capacha monochrome, bule sherd, undecorated (not illustrated). Surface, monero discard.

9807 a, b. Capacha monochrome, painted variants; respectively pinched-rim bowl and olla(?) sherds (Figs. 28 *e; 31 e, f*). Surface, monero discards.

9811. Burned clay with cane imprint. General digging; association Capacha.

9815. Trade sherd, excised, curvilinear design (Fig. 31 *k*). General digging; association mixed.

9961. Capacha monochrome; sherd, better quality than usual (Fig. 31 *j*). Vicinity of Burial 3.

B-3. La Parranda (Site A)

9617. Capacha monochrome, bule, decorated (Figs. 7, no. *12; 20 a*). General digging.

9617 a. Capacha monochrome, bule, decorated (Fig. 7, no. *13*). General digging.

9621. Capacha monochrome, miniature vessel (Fig. 14 *e*). General digging.

9809. Capacha monochrome, painted variant; bowl, flaring rim (Fig. 28 *d*); restorable. General digging.

9836. Capacha monochrome, bule, decorated (not illustrated). Surface, monero discard.

9846 a–c. Obsidian chips. Test 6. Association mixed, but chiefly Capacha.

B-4. Terreno de Jesús Gutiérrez

8189. Capacha monochrome, bule, decorated (not illustrated); semirestorable. Surface, monero discard.

8190. Capacha monochrome, bule, decorated (not illustrated); semirestorable. Surface, monero discard.

8191. Capacha monochrome, bule, decorated (not illustrated); semirestorable. Surface, monero discard.

B-5. Parcela de Luis Salazar

10244. Capacha monochrome, bule sherd, decorated (not illustrated); not restorable. Vicinity of Burial 1, but not in direct association.

10245. Capacha monochrome, bule sherd, decorated (not illustrated); not restorable. Same provenience as preceding.

10247. Figurine leg (Fig. 33 *d*). Same provenience as preceding. Association probably Capacha.

10248. Mano fragment, plano-convex (not illustrated). Same provenience as preceding. Association probably Capacha.

10249. Grinding-stone fragment, once used for red paint (not illustrated). Same provenience as preceding. Association probably Capacha.

B-6. La Capacha

9651. Capacha monochrome, bule, decorated (Figs. 9 *j; 16 d*). General digging.

9652. Capacha monochrome, small cántaro, decorated (Figs. 10 *f; 22 a*); semirestorable. General digging.

9653. Stone "plate" (Fig. 42 *b*). General digging, in close proximity to 9652 and 9654.

9654. Stone "spoon" or "ladle" (Figs. 42 *a; 43 c, cc*). General digging, in close proximity to 9652 and 9653.

B-7. El Barrigón

8888. Capacha monochrome, cántaro with subconic neck (Fig. 10 *d*); not restorable. General digging.

8890. Stone-ax fragment (not illustrated). Test 6; association mixed and Capacha affiliation not secure.

8891. Worked-stone fragment; corner of an unidentified artifact (not illustrated). Provenience and association same as preceding.

8892. Obsidian chip. Test 7; association firmly Capacha.

9253. Capacha monochrome; painted variant, purple-black and rose red, combined with incision (Fig. 31 *g*); sherd. Surface, monero discard. A similar sherd is of the same provenience.

B-8. Terreno de Fidel Valladares

This cemetery produced no burials identifiable as Capacha through accompanying offerings. Nevertheless, one entire and one restorable specimen, both Capacha, come from this burial ground and, except for a few fragments of Incised unclassified and Red unclassified, sherd material is overwhelmingly Capacha. The Incised fragments are small—thin, with broad-line incision; they do not resemble ceramics known in non-Capacha association and perhaps may prove to constitute an unrecognized Capacha ingredient. However, the Red unclassified is an orange-red which tends to occur in much later association, starting with the Colima phase. With these reservations, the specimens listed below may be considered Capacha until fuller data are available.

9993. Bead of green stone (Fig. 35 *f*); small (5 mm diameter), roughly disk shaped. Infant Burial 1, no other furniture; probably Capacha.

9995. Grinding-stone fragment (Fig. 37 *c*). Burial 3, no ceramics. Probably Capacha.

9996 a, b. Manos (Fig. 40 *g*, *f*). Same provenience as preceding.

9998. "Nutting" stone (Fig. 41 *b*). Surface.

9999 a, b. Grinding-stone fragments (9999 b, Fig. 38 *d*). Surface.

10218. Capacha monochrome, olla fragment (Fig. 11 *e*). Surface and general digging.

10219 a. Trade sherd, excised (not illustrated). Surface.

10221. Ax fragment, three-quarter groove (not illustrated). Same provenience as preceding.

10222. Worked stone object, notched (Fig. 45 *b*); not classified. Same provenience as preceding.

10223. Obsidian chip. Same provenience as preceding.

10226. Worked sherd, rectangular (Fig. 31 *s*). Test 12; association predominantly but not exclusively Capacha.

10227. Red pigment, presumably hematite; small lump. Same provenience as preceding.

10228. Pebble, symmetrical, probably an artifact (not illustrated). Same provenience as preceding.

10231 a–c. Obsidian chips. Test 15; association chiefly Capacha.

10232. Mano (Fig. 40 *h*), with shallow pit on both faces. Test 15; association strongly Capacha.

10235. "Nutting" stone, pitted on both faces (Fig. 41 *c*). Tests 1–3, lumped; association Capacha.

10236 a. Obsidian scraper (Fig. 36 *d*). Same provenience and association as preceding.

10236 b–d. Obsidian chips. Same provenience and association as preceding.

10239. Worked sherd, small, roughly circular (Fig. 31 *q*). Test 7; association overwhelmingly Capacha.

10240. Tortoise-shell(?) fragment (not illustrated). Same provenience and association as preceding.

10241 a, b. Obsidian chips. Provenience and association same as preceding.

10363. Rubbing stone, biscuit shaped (not illustrated). General digging.

B-10. Quintero

This cemetery was extensively pitted by moneros, but several productive tests were fitted into untouched areas between their excavations and on the periphery of the site. Surface and general digging produced mixed sherds, hence material listed below of such provenience is not securely Capacha, although cemetery provenience is certain. Except for two large sherds, rose colored, of Red unclassified —presumably relics of a post-Capacha interment flush with the surface and otherwise carried away by the adjacent arroyo—all burials identifiable as to phase were Capacha. Two vessels, attributed respectively to Burials 8 and 9, might better have been considered products of general digging; burial numbers were assigned in anticipation of accompanying skeletal material, but no bone was found in association. In addition, several interments, presumably Capacha (Table 2), were not accompanied by offerings, hence phase allocation is not secure.

9739. Capacha monochrome, small olla, undecorated (Fig. 11 *f*). General digging.

9740. Capacha monochrome, rose-slipped variant; olla, decorated (Figs. 28 *a*; 29 *b*). General digging.

9749. Capacha monochrome, incurved bowl, decorated (Figs. 12 *h*; 22 *f*; 23 *b*). General digging.

9750. Obsidian chip. Adjacent to 9749; probably Capacha.

9784. Grinding-stone fragment (Fig. 37 *b*). General digging.

9785. Mano (Fig. 40 *a*). General digging.

9908. Stone object, palette shaped; wall with horizontal groove (Fig. 45 *a*). Surface.

9909. "Nutting" stone (Fig. 41 *a*). Surface.

9910 a, b. Mano fragments (Fig. 40 *b*, *c*). Surface.

9911. Stone animal foot(?) (Fig. 45 *c*). Surface.

9912 a. Pebble, water worn; two faces show use as polisher (Fig. 44 *c*). Surface.

9912 b. Pebble, unusual color; no sign of use and may not be an artifact (not illustrated). Surface.

9917 a. Capacha monochrome, possibly bottle neck (Fig. 10 *h*); sherd. General digging.

9926 a–c. Worked sherds (Fig. 31 *p*, *o*, *r*; *r* may not be Capacha). General digging.

10216 a. Capacha monochrome; miniature vessel fragment, decorated (Fig. 14 *f*); semirestorable. Surface, presumably monero discard.

10314. Grinding-stone fragment (Fig. 38 *b*). Surface.

10344. Amulet(?) fragment, of black, nonvesicular basalt (Fig. 44 *b*). Vicinity of Burials 7 and 9; association predominantly Capacha.

Unnumbered site, Mesa del Salate, Los Ortices, Colima

This is a cemetery thoroughly looted by moneros. The few surface sherds indicate mixed affiliation. Of this surface material, two fragments indicate the presence of at least one Capacha interment. Data are too few to warrant listing this as a Capacha cemetery.

6321. Capacha monochrome, cántaro sherd with continuous decoration forming a lattice or network (Figs. 24 *a*; 26 *a*). Surface, monero discard.

6322. Capacha monochrome, rose-slipped variant; cántaro or olla fragment (indeterminate) with raised ribs (Fig. 29 *i*). Same provenience as preceding.

C. ALLEGED ASSOCIATION IN GRAVE LOTS

Listing of specimens in this category rests on the authority of moneros, and it seems eminently sensible to take advantage of all information that can be salvaged from them. Grave-lot association is of no economic interest to any looter, hence there is no reason for deception. Indeed, internal evidence is very much in favor of accurate reporting by moneros, and I am inclined to accept their identification of grave lots, with the reservations mentioned below, under D. All material listed under C is purchased.

C-6. La Capacha

Said to be from monero pit no. 1, "compartment" 1

8679. Capacha monochrome, cántaro, with faint channel at base of neck (Fig. 10 *g*).

8680 a, b. Pottery ocarinas (Fig. 35 *a, b*).

8681. Stone dish, tripod (Figs. 42 *d*; 43 *a, aa*).

Said to be from same monero pit no. 1, "compartment" 2

8682. Grinding stone, miniature, four nubbin feet (Figs. 39; 43 *d*); no pottery in direct association.

8683 a–f. Small stones and pebbles (not illustrated).

Said to be from same monero pit no. 1, "compartment" 3

8684. Capacha monochrome, bule, decorated (Figs. 9 *k*; 16 *e*; 17 *f*).

8685 a, b. Capacha monochrome, miniature effigy dishes (Figs. 14 *a, b*).

8686. Stone dish, oval, two cusped feet (Figs. 42 *e*; 43 *b, bb*).

Said to be from same monero pit no. 1; informant uncertain of individual "compartment"

8687. Stone ax, three-quarter groove (Fig. 44 *a*).

Said to have been found together in same Capacha cemetery; no mention of skeletal remains

8709. Pottery pendant, animal effigy (Fig. 35 *d*).

8710. Capacha monochrome, miniature effigy dish, tetrapod (Fig. 14 *c*).

Said to come from a single grave lot in same Capacha cemetery

8712. Capacha monochrome, bule, decorated (Figs. 9 *c*; 16 *f*; 17 *h*).

8713. Capacha monochrome, cántaro with raised ribs (Figs. 10 *a*; 26 *c*; 27 *a*).

8714. Capacha monochrome, stirrup pot, variant with three tubes (Figs. 13 *e*; 25 *b*).

D. ALLEGED ASSOCIATION IN MONERO PITS

Here, again, moneros have provided the information, and there is far greater chance of error, unintentional though it be. Moneros blandly disregard skeletal material because it has no commercial value; moreover, they undercut every pit, so that material from one of their "excavations" well may come from more than one interment. It may be assumed that given provenience is accurate as far as cemetery association is concerned but is not necessarily reliable in terms of individual grave lots.

All items under "D" refer to La Cañada, Comala. Specimens 9624; 9625 a, b; 9626; 9627; and 9640 actually were discards, whose fragments we collected on the surface; they were recognized by the moneros who thereupon provided pit provenience. Other specimens were purchased.

D-2. La Cañada

Said to be from monero pit A

9632. Capacha monochrome, compound plate (Figs. 13 *b*; 26 *e*).

9633 a. Capacha monochrome, bowl, flaring rim (Fig. 12 *l*); unusually poor quality.

9633 b. Unclassified ware; small vessel with partial rose wash, simulating red on brown(?), finger grooving (Figs. 28 *b*; 29 *a*).

In addition, it was reported that a bule and an unfired figurine came from this same monero pit.

Said to be from monero pit B

9634. Capacha monochrome, cántaro, rim incomplete (Fig. 10 *c*).

9635. Capacha monochrome, olla (Fig. 11 *c*); poor quality.

9636. Capacha monochrome, bowl, flaring rim (Fig. 12 *m*); poor quality.

In addition, moneros reported having found one bule and a small unfired figurine, with red paint, in this same pit.

Said to be from monero pit C

9624. Capacha monochrome, painted variant; tortoise effigy (Fig. 30 *c, cc*).

9626. Capacha monochrome, painted variant (Fig. 30 *b, bb*).

9637. Capacha bule, decorated (Figs. 9 *b*; 15 *b*; 17 *c*); unusually good quality.

Several additional bules allegedly were removed from this same pit.

9638. Cranium, collected on surface (Appendix V, no. 9638).

Said to be from monero pit D

9625 b. Capacha figurine fragments, semifired (Figs. 32 *b*; 33 *e*).

9639. Capacha monochrome, olla (Fig. 11 *a*).

Four bules were attributed to this same pit.

Said to be from monero pit E

9625 a. Figurine fragment, not Capacha, but of a style associated with Ortices-Comala phases (not illustrated). Presumably informant's error.

9627. Capacha monochrome, bule, undecorated (not illustrated). Restorable.

Said to be from monero pit F

9640. Capacha monochrome, form best considered an olla (not illustrated).

Several bules and "unfired" figurines allegedly were removed from this same pit.

E. ALLEGED CEMETERY ASSOCIATION

Again, moneros have provided the information for this section, and there seems no reason to doubt it.

E-2. La Cañada

9641 a. Capacha monochrome(?), painted variant; olla, red on brown (not illustrated). Purchased with two other vessels, neither of which is Capacha, although present specimen might be an aberrant example.

9648. Capacha monochrome, bule (not illustrated). Gift.

9649 a, b. Capacha figurines, "unfired" (Fig. 34 *a–aaaa, b–bbbb*), but said to be better baked than most. Repossessed by monero from dealer to whom they had been sold; subsequently purchased by me.

E-3. La Parranda (Site A)

9631. Capacha monochrome, bule (not illustrated). Purchased.

E-6. La Capacha

8715 a, b. Capacha monochrome, bules (8715 a, **Figs.** 9 *i*; 20 *f*; 8715 b, Fig. 20 *e*). Said to have been found in general vicinity of 8712–8714 (see C-6), but not directly associated with the same burial. Purchased.

8716. Capacha monochrome, basin-cauldron (Fig. 12 *b*). Same provenience as preceding.

8717. Capacha monochrome, fragment of horizontally compound cántaros (Fig. 13 *c*). Same provenience as preceding.

8718. "Nutting stone" (Fig. 43 *e*). Same provenience as preceding.

8719 a, b. Pottery beads(?), small, funnel shaped (8719 a, Fig. 35 *c*). Same provenience as preceding.

8720 a–j. Obsidian chips. Same provenience as preceding.

8720 k. Obsidian chip, apparently somewhat retouched. Same provenience as preceding. Alleged Capacha association of 8720 a–k seemingly confirmed by comparatively early hydration dates (Chapter 3: *Absolute Dating*).

8807. Capacha monochrome, bule (Fig. 20 *b*). Purchased; cemetery provenience only.

E-8 Terreno de Fidel Valladares

Specimens here listed were acquired by gift or by purchase.

9970. Capacha monochrome, stirrup pot, variant with three tubes (Figs. 24 *b*; 25 *c*).

9971. Capacha monochrome, incurved-bowl sherd (Fig. 12 *g*).

9973 a, b. Palettes(?) of limestone (Fig. 43 *f, g*).

9994. Pendant of rock crystal; slender, pestle shaped (Fig. 35 *e*).

9997. Stone dish (Fig. 42 *c*).

10138. Capacha monochrome, cántaro with ridged decoration (Fig. 26 *d*).

E-10. Quintero

All specimens listed below were purchased.

9684. Capacha monochrome, bowl, slightly incurved at rim (Fig. 12 *e*).

9685. Capacha monochrome, bowl, incurved, pinched rim (Fig. 12 *i*).

9686. Capacha monochrome, bowl, incurved, perforated for suspension (Fig. 12 *f*).

9687. Capacha monochrome, cántaro, raised ribs (Figs. 10 *b*; 27 *b*).

9688. Capacha monochrome, olla (Fig. 11 *d*); poor quality.

9689. Capacha monochrome, painted variant; olla, red on brown (Fig. 30 *a*).

9690. Capacha monochrome, bule; rim missing (Fig. 15 *c*).

9692. Grinding stone (Fig. 38 *c*).

10016. Grinding stone (Fig. 38 *a*).

10017. Mano (Fig. 40 *d*); said to have accompanied 10016.

10018. Mano (Fig. 40 *e*).

10019. Capacha monochrome, plate; nubbins at opposite sides of rim (Fig. 12 *n*).

10020. Capacha monochrome, squat olla, decorated (Fig. 22 *c*).

10021. Capacha monochrome, hemispherical-bowl sherd; squared rim, with punches (Figs. 12 *d*; 22 *d*).

10022. Capacha monochrome, bowl, incurved (Fig. 12 *j*).

10023. Capacha monochrome, bowl, incurved, with eight perforations at rim (Fig. 12 *k*).

10024. Capacha monochrome; fragment of horizontally compound cántaros (Fig. 13 *a*).

10025. Capacha monochrome, cántaro (Fig. 10 *e*).

10026. Capacha monochrome, bule, undecorated (Fig. 9 *f*).

10027. Capacha monochrome, bule, decorated (Fig. 17 *g*).

10028. Capacha monochrome, bule, unusual decoration (Fig. 20 *d*). Barely restorable. Part purchased; completed with sherds collected on surface from monero discards.

10029. Capacha monochrome, bule, decorated (Fig. 20 *c*); semirestorable.

10323. Capacha figurine (Fig. 34 *c–ccc*). Purchased from Colima resident to whom it had been sold by monero who excavated at Quintero.

Unnumbered site, Potrero El Terrero, near El Diezmo, Colima

There is no information concerning this presumed Capacha cemetery other than the purchased specimen listed below. The vendor refused to give more precise provenience.

5533. Capacha monochrome, bule, decorated (Figs. 9 *a*; 17 *a*; 19 *d*).

Appendix V
SKELETAL REMAINS

José Antonio Pompa
Departamento de Antropología Física, INAH

This report concerns some of the characteristics of the skeletal material associated with the Capacha phase in the state of Colima, Mexico, and believed to date to approximately 1450 B.C., adjusted to 1870–1720 B.C.

Because of the poor state of preservation, considerable reconstruction and consolidation were necessary but, even so, no piece can be considered complete. Owing, however, to its comparative antiquity and its cultural associations, the material is welcome. It contributes to our knowledge of human remains from west Mexico and provides clues to possible connections with other areas of the Americas.

The sample consists of seven incomplete crania and five incomplete mandibles, as well as fragments of an axis, a scapula, two femora, two tibia, and a humerus.

8632

Site. El Barrigón, Buenavista, Cuauhtemoc, unnumbered burial (Appendix IV-A-7, 8632).

Preservation. Poor.

Age. Adult.

Sex. Male.

Intentional cranial deformation. Tabula erecta fronto-occipital, pseudo-annular variety (Fig. 46 *a*).

Material. An incomplete skull (Figs. 46, 47), reconstructed and partially restored. The base is missing (Fig. 46 *b*), as well as the left facial portion and the nose bones (Fig. 46 *c*). The left temporal is disarticulated. The skull is robust and the muscular attachments are very well marked on the occipital bone. The thickness of the cranial vault is marked, especially in the bregmatic area, with the cranio-metric point (bregma) located in a fossa (Fig. 47 *a*). The sutures are very much obliterated, especially the sagittal one. The lambdoid suture shows the apical and wormian bones, with the inion prominent. The supraorbital arch is clearly marked (Figs. 46 *c*; 47 *b*). The upper central incisors are shovel shaped and show advanced attrition.

An infectious process evident in the vicinity of the upper right canine root (Fig. 46 *c*) has resulted in destruction of this zone in the maxilla. The right zygomatic bone shows exostosis in the central part of the external surface, and the articular surface of the glenoid cavity of the right temporal bone (Fig. 47 *b, c*) likewise exhibits exostosis. There are two small fossae (Fig. 47 *c*), probably caused by ante-mortem traumatisms, one on the rear portion of the left parietal, the other on the frontal bone, just above the right orbital arch, toward the zygomatic process of the frontal bone.

9638

Site. La Cañada, Comala, surface, monero discard (Appendix IV-D-2, 9638).

Preservation. Poor.

Age. Adult.

Sex. Male.

Intentional cranial deformation. Tabula erecta fronto-occipital, but not very clear.

Material. An incomplete skull and an incomplete mandible. Of the skull, we have part of the frontal bone (the squama); the occipital (without the basis), articulated with the posterior portion of both parietals; part of the mastoid area and of the left temporal, plus the mastoid and petrous regions of the right temporal. The thickness of the cranial vault and the muscular attachments are clearly visible, as is the superciliary arch. The wormiams at the lambdoid suture were lost post-mortem.

We have part of the body of the mandible, with the first right molar in place; it shows advanced attrition. The two central lower incisors were lost ante-mortem, as was the second lower right molar. The remaining dental pieces were lost after death, but two premolars, one molar, and a root were recovered, although not in situ. There is no indication of pathological condition.

9752

Site. Quintero, Ixtlahuacán, Burial 5 (Appendix IV-A-10, 9752).

Preservation. Poor.

Age. 4–6 years.

Sex. Indeterminate.

Intentional cranial deformation. Tabula erecta fronto-occipital.

Material. Skull and mandible are broken and incomplete. The edges of the fractures are eroded, for which reason only the posterior part of the vault was reconstructed. The maxilla and the body of the mandible are in a compacted block of earth. No pathological condition is apparent.

10293 a, b

Site. La Cañada, Comala, Burial 5 (Appendix IV-A-2, 10293 a, b).

Preservation. Poor.

Age. Two adults.

Sex. Indeterminate.

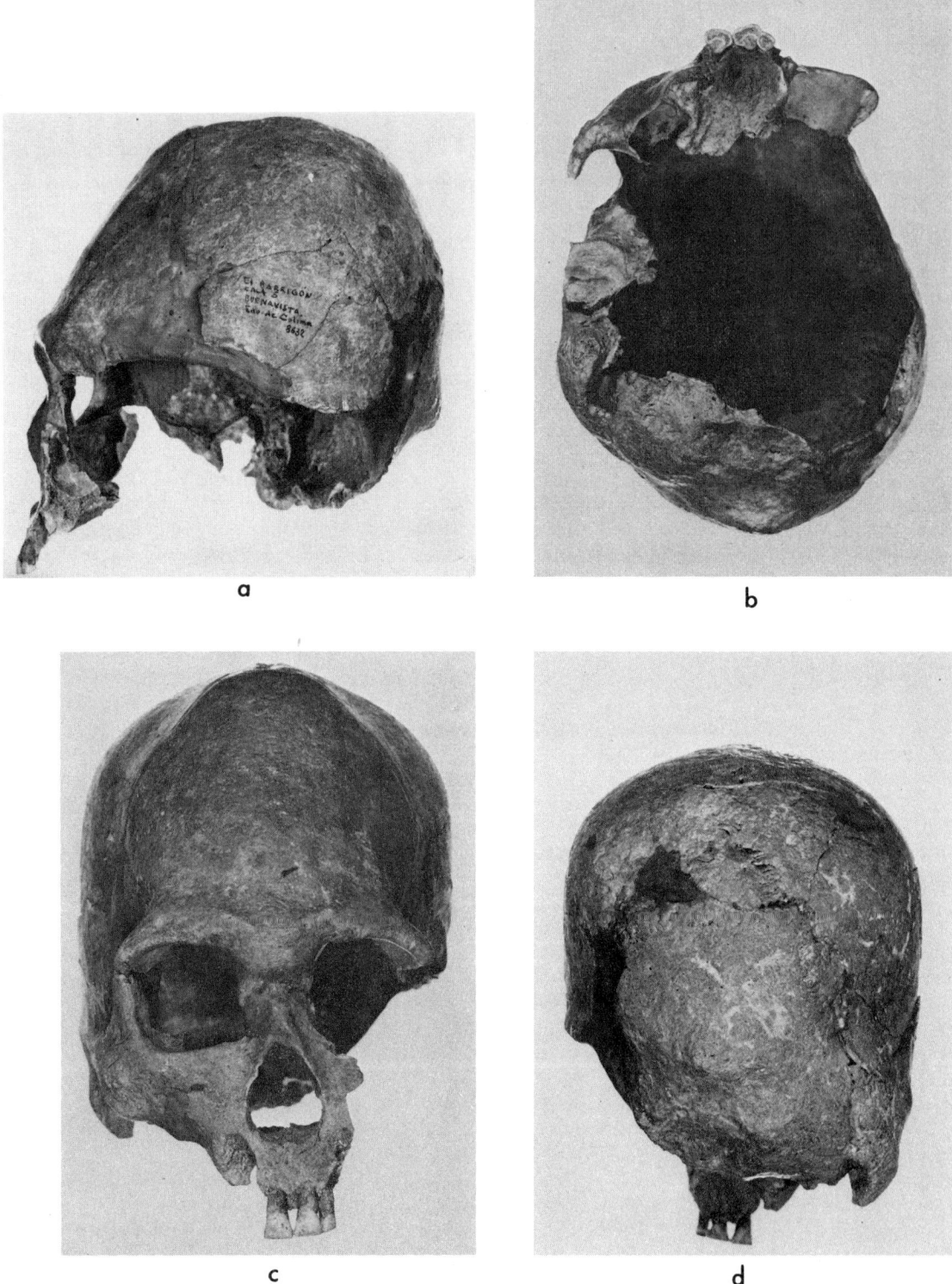

Fig. 46. Capacha-phase cranium, Appendix V, no. 8632. Unnumbered burial, El Barrigón, Buenavista. See also pp. 18, 48, 92, 97.

Intentional cranial deformation. One skull, 10293 a, shows slight evidence of tabula erecta, pseudo-annular.

Material. Two incomplete crania, one mandible, and fragments of long bones, all found within a bule (Appendix IV-A-2, 9722). Despite the poor condition, it was possible to reconstruct the left portion of skull 10293 a.

a

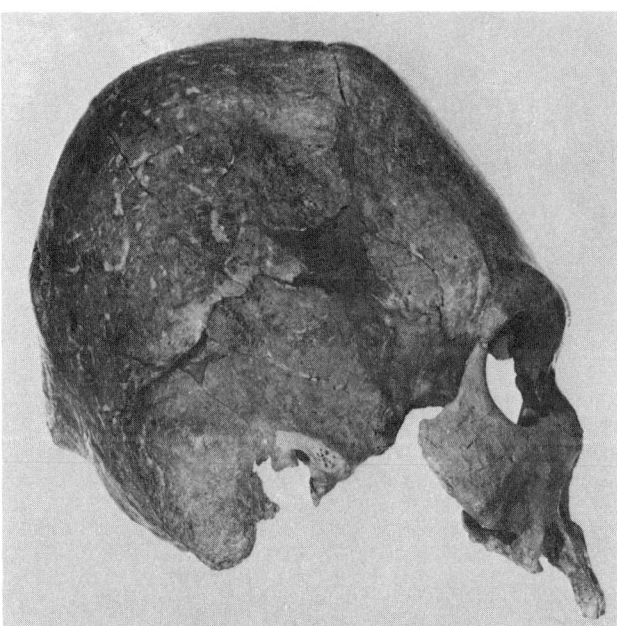

b

c

Fig. 47. Capacha-phase cranium, Appendix V, no 8632. Unnumbered burial, El Barrigón, Buenavista. See pp. 18, 48, 92, 97.

The bones of the face and of the base are missing, and the left temporal could not be articulated. There is no sign of pathological lesion, but slight post-mortem deformation is appreciable.

Of skull 10293 b, we have only the posterior part of the two articulated parietals, hence it is not possible to determine if there were intentional deformation.

The left ramus of the mandible is missing. The first two molars were lost ante-mortem; the second and third molars are in place and show advanced attrition. The remaining dentition was lost post-mortem.

No parts of the post-cranial skeleton can be identified with either skull, but there are fragments of the shafts of the two tibia and of both femora, in addition to small parts of the same shafts and of both skulls.

It may be assumed that these two individuals represent a secondary interment within the pottery container.

10300

Site. Quintero, Ixtlahuacán, Burial 7 (Appendix IV-A-10, 10300).

Preservation. Poor.

Age. Adult.

Sex. Indeterminate.

Intentional cranial deformation. Not perceptible.

Material. Part of the right parietal is destroyed. The external surface of the vault and the edges of its fractures are eroded. We have the two palatine processes of the maxilla and fragments of the mandible body, as well as small fragments of the skull.

10305

Site. Quintero, Ixtlahuacán, Burial 10 (Appendix IV-A-10, 10305).

Preservation. Poor.

Age. Adult.

Sex. Female.

Intentional cranial deformation. Not perceptible.

Material. An incomplete skull and incomplete mandible, plus fragments of an axis, a scapula, and a humerus.

Of the skull, we have the scamous portion of the frontal bone, broken on the edges and semiarticulated with both parietals. There are also fragments of the disarticulated maxillae, as well as small pieces of the cranial vault.

The mandible is small and hardly robust. A large portion of the left ramus is missing. Dental pieces lost post-mortem are the first left lower molar, the third right lower molar, and the first right upper molar. The central superior incisors are shovel shaped, and all the dentition shows marked attrition.

Of the axis, we have only a fragment of the right side, which has the articular surface for the atlas, the transverse foramina for the vertebral arteria, and part of the lamina. The scapula is from the right side and has only the spine, without the acromial process.

The humerus also is from the right side; the proximal epiphysis is missing, as are the capitulum and medial epicondyle. The humerus shows olecranon foramina; it is rather slight and almost without torsion.

In summary, the Capacha skeletal material pertains to at least seven individuals, of which six (8632; 9638; 10293 a, b; 10300; 10305) are adult and one (9752) a child. There are two males (8632, 9638); one female (10305); and four individuals of indeterminate sex, including the child (9752; 10293 a, b; 10300).

Of the seven skulls, four (8632, 9638, 9752, 10293 a) are intentionally deformed. The two males (8632, 9638) exhibit the tabula erecta fronto-occipital type, one of which is the pseudo-annular variety. One of the individuals of indeterminate sex (10293 a) has tabula erecta fronto-occipital deformation, likewise pseudo-annular. In addition, the child's skull shows tabula erecta deformation.

This specific type of cranial deformation is of considerable antiquity in Mexico, as attested by a skull in preceramic association from the Texcal cave, in Puebla (Romano 1972). It also is preceramic on the north and central coasts of Peru and is reported for the Machalilla phase of the Ecuadorean coast (Munizaga 1965: 230). Irrespective of the dating of Machalilla (see Chapter 4: *The Chronological Quandary*) tabula erecta deformation is common to that phase; to Capacha, Colima; to El Opeño, Michoacán; and to Tlatilco, Estado de México.

With respect to the dentition of the Capacha sample, attrition of the masticatory surface is marked, as is the shovel shape of the upper central incisors. No case of dental mutilation was observed.

Unfortunately, owing to the small size of the sample and the poor state of preservation, it has not been possible to make metrical or morphological comparisons, for which reason the present report is limited to the descriptive aspects. However, it may be noted that, on the whole, the cranial vaults—including even the child's skull—are consistently thick, with a range from 5 to 11 mm.

NOTES

1. Most of Colima's rural adult male population, and a few of the women and adolescents as well, are active moneros. As one man says, with enormous dignity, "Our profession is that of excavating."

2. Originally, this summary was much longer and included a discussion of Colima in the context of west Mexico and of Mesoamerica. It turned out to be too voluminous to serve as an introduction to the Capacha phase, so some of the material has been reserved for a general overview of Colima archaeology, to be prepared later, once the ceramic sequence in east Colima is in order. Any general discussion of Colima archaeology should be well illustrated, but if figures were to accompany the present summary, the Capacha material would be eclipsed. Therefore, this introduction is greatly compressed and includes no illustrations, but references to key specimens published elsewhere are provided.

3. Amoles wiped ware has been called La Loma wiped and appears under that name in Meighan, Findlow, and De Atley (1974). However, years before, a quite unrelated ware in the Tuxcacuesco area was designated as La Loma red on brown (Kelly 1949: Pl. 16 *a–d*). To avoid confusion, the shadow-striped ware shared by the Ortices and Comala phases is now being called Amoles wiped.

4. This agrees with information from the Morett site, where the "level above that containing this sherd [of Ortices polychrome] is dated by radiocarbon at 2,025 years ago . . . It is illustrated in plate 21, *h*" (Meighan 1972: 52). The problem here is that, apparently through editorial error, the Ortices sherd appears in his Plate 21 *i,* not *h*).

5. But by no means all. For example, certain handsome, hollow specimens (Médioni and Pinto 1941: frontis., nos. 142, 155; von Winning 1969: no. 57; and Kan, Meighan, and Nicholson 1970: no. 130) have been found near Chanchopa and elsewhere in eastern and southeastern Colima. Aveleyra (1964) publishes the photograph of a similar specimen attributed to Tecomán, which might mean Chanchopa (reference to this photograph is difficult, for neither page nor plate is numbered; it seems to be Plate 111). Unfortunately, this style of effigy is not identified as to phase, and various other large, hollow figures are without phase affiliation.

6. Owing to re-use of tombs (Kelly 1978), dates based on organic material from burial chambers are difficult to allocate to phase. Taylor (1970: 165, Table 1, UCLA 1438) has published a corrected date of 90 B.C., derived from shell from a Colima shaft tomb; he cites Berger and Libby (1970), a paper I have been unable to locate. It is quite likely that the tomb in question may belong to the Comala phase, but there are other possibilities. In the Armería axis, shaft tombs produce wares of the Ortices and Comala phases, and at least one contained ceramics being designated as Manchón, apparently intrusive in the area (Kelly 1978). Furthermore, it appears that the unplaced Parranda phase (see Chapter 2: *Comments*) will prove to be associated with tombs, and the same may hold for Capacha. In short, there are five possibilities, without considering east and southeast Colima where, at the very least, four tentatively defined phases are associated with tombs.

7. The serpent with a head at either end was known to the Mexica as Maquizcoatl, or "bracelet snake." As such, it appears in the Florentine Codex (Dibble and Anderson 1963: 79; reference courtesy of Mrs. Thelma Sullivan) as a potential augury of death (cf. Furst 1965b: 20; Kelley and Kelley 1971: 117–19). Accordingly—if an immense break in time, space, and probably culture be permitted—the double-headed snake might be a fitting motif for funerary ware.

In Colima, the symbol appears in relief on the cylindrical vessels mentioned above, but my impression is that it also occurs, at least on occasion, three-dimensionally on the so-called "baskets," which are effigy incense burners. Unfortunately, published photographs of such specimens (for example, von Winning 1969: no. 116/117; Kan, Meighan, and Nicholson 1970: no. 199; Schöndube 1974: Fig. 10 *b*) do not show the double-serpent head clearly. Nevertheless, a variant, attributed to the eastern slopes of the Volcán Nevado de Colima certainly depicts a two-headed snake (Schöndube 1974: Fig. 6). I have no direct information concerning the ceramic phase of such specimens, but moneros state that they are found in tombs; this rather suggests Comala, which also is the phase of the cylindrical vessels.

Furst (1965b: 20) calls attention to the prevalance of the double-headed snake in Mexico and Peru, and a detailed comparative study including representation, associations, and chronological position might be rewarding.

8. After the above was written, Prof. Luis Javier Galván saw some of the material and suggested a significant resemblance between the multiple-brush decoration of Mazapan ware and that of the cántaros of the Colima-Armería phases. Mazapan includes no cántaros with such ornament, but the latter occurs on the floor of the red on cream plates of Linné's first category. Apart from the basic difference in vessel form, Mazapan multiple-brush decoration is very much simpler than is the usually complex Colima-Armería ornament (Kelly 1978: Fig. 24).

9. The current work of Dr. Phil C. Weigand in the vicinity of Eztatlán has amplified and sharpened the chronology for that area. He retains Huistla as the phase name for the early Postclassic and indicates (personal communication, 30 July 1977) that this is the time slot where "Mazapan figurines appear in abundance."

10. Pendergast (1962b) has prepared a very useful classification for Mesoamerican metal artifacts. It should now be expanded to include specimens at present on exhibit and in storage in the Museo Nacional de Antropología e Historia. Many of the Colima specimens fit the Pendergast scheme; some do not. In any event, the Colima material here mentioned, plus some copper objects purchased a few years ago from a monero who had "worked" in a late (Tolimán-phase) cemetery at Chachahuatlán, Tuxcacuesco (Jalisco), constitute a welcome addition to our knowledge of metal working in west Mexico. Apart from some forms that were not known to Pendergast, the lot contains a few examples that will change some of his distributions. Obviously, these data cannot be included in the present summary and will have to be treated elsewhere.

11. I made a considerable search, in the hope of finding a trace of the famous sixteenth-century settlement of Alima (Lebron 1951: 12, 21; Sauer 1948: Map 3; Brand and others 1960: 145, Table 6); it may no longer exist, for large areas of the lower Coahuayana valley literally were carried away during the 1959 cyclone, and some parts were stripped of all surface soil.

12. Lebrón (1951: 64–65) makes a significant comment to the effect that, in his day, the Villa de Colima occupied the site where formerly was situated the Indian pueblo of Tuxpan (now in Jalisco). One would think this statement might be subject to confirmation archaeologically, yet the data are not conclusive. Painted ware generically related to Autlán polychrome (of Chanal association in Colima) is reported for the Tuxpan-Tamazula area (Schöndube 1973–74 I: 80–83), but similar pottery extends at least as far north as the Sayula basin. The distribution of the less distinctive "rustic" Chanal ceramics might be explored from this point of view.

13. Illustrated in Kelly (1949: Fig. 38). Probably a comal variant; several sherds suggest a vessel too large to be used as a dipper. Some fragments evidently had an arched handle, but precise form and placement are uncertain. No entire specimen is known. As a guess, this slant-mouthed vessel might have been handy for toasting chilies, or seeds such as maize, squash, chili, and cacao—operations which in modern rural Mexico are done on the comal. The raised rim along part of the periphery would keep the contents from scattering.

14. Although not necessarily so. A careful, systematic study of tripod bowls should be revealing. It should include many aspects, such as the proportions and shape of the bowl; the form, proportions, and placement of the feet; decoration; and floor treatment. Such details should lend themselves admirably to the computer approach so popular these days.

15. Those that were removed from the best-known "pyramid"—that of East Chanal, which was "excavated" years ago—have been recovered by the Museo Regional de Colima. A careful study of motifs and of style should be enlightening.

16. In connection with such a postulated movement, one ponders the possibility of a link with Otomí speech reported for parts of Jalisco and Nayarit ([Cortés] 1937), combined in some pueblos with a form of "mexicano." Prof. Leonardo Manrique (personal communication, 1974) feels that if the Otomí language really is involved, it probably moved into the west in comparatively recent centuries before the Spanish Conquest.
No contemporary vocabulary seems to have survived, and there is no known indication of non-Nahual speech in the Mi-

choacán-Colima-Jalisco stretch in recent decades. In 1940–41, I recorded a brief Nahual vocabulary in Ayotitlán (Jalisco), and in 1941, at the suggestion of Dr. Carl Sauer, Mr. Jean B. Johnson made an extensive linguistic survey. His unpublished vocabularies, now in the possession of Mrs. Irmgard Weitlaner Johnson, cover the considerable stretch from Coire (Michoacán) to Ayotitlán (Jalisco) and include Ixtlahuacán and Suchitlán (both, Colima) and Tuxpan (Jalisco).
We may never know whether the language reported in the Cortés document is actually related to that which today is called Otomí. Clearly, it was something other than Nahual, for there is specific mention of "the language of Mexico" and of *naguatlatos*. Here, Otomí is not used (as it was in the vicinity of Mocorito, in Sinaloa) simply to designate a "corrupt and barbarous" language, because Otomí is "the language that in the Province of Mexico is considered as most difficult" (Sauer 1934: 29).
The Cortés document is the earliest (and almost the only) source of ethnographic and linguistic information on the area between the "city" of Autlán (Jalisco) and Etzatlán (on the Jalisco-Nayarit border). It contains fascinating material concerning dual chieftainship, with one *calpixque* Otomí and the other Nahual in some pueblos. It also gives data, pueblo by pueblo, with respect to ethnic boundaries, local wars, village sites, barrio divisions, wells for potable water, house types, agriculture, crops, irrigation, and markets and commerce.

17. There seems no need for hesitancy in postulating movement of peoples. Recently, I have read anew some of the chronicles, as well as Carrasco (1971) and Davies (1973). The population of central Mexico evidently was extraordinarily mobile in the centuries just prior to the Spanish Conquest, and, without particularly compelling reasons, groups shifted considerable distances. The heterogeneous composition reported in early accounts for many settlements in the central valleys confirms the traditions of physical mobility. Moreover, the sixteenth-century pueblos of Colima were anything but uniform in speech and culture (Lebrón 1951: 11, 14).

18. Concentrated in Colima and Jalisco: at the mouth of the Río Coahuayana; at El Chanal; in the Autlán-Tuxcacuesco-Zapotitlán stretch upstream, in the Armería drainage; in the adjacent Tapalpa highlands; eastward, in the Sayula and Zapotlán valleys; plus a weak extension northwest, to Mascota.
The relationship of Autlán polychrome and its variants to the engraved and painted tripod bowls of the Chapala area and elsewhere awaits definition.

19. Since the above was written, the Amapa report (Meighan 1976) has appeared, and it has been added to the list of works cited herein. Statements concerning influences from central Mexico in Postclassic times are essentially those outlined earlier by Meighan (1974).

20. After the present report had been submitted, Greengo and Meighan (1976) published a short paper with illustrations of a collection of Capacha vessels reputedly from Apulco, near Tuxcacuesco, Jalisco, not far from the Colima border. This occurrence does not alter materially the known distribution of Capacha wares, and the specimens provide a welcome addition to our knowledge of Capacha ceramics.
Included are two bules, an effigy cántaro, and various miniature vessels of tumbler form; several of the latter are cinctured, with raised ridges bounding the restriction. Of the six stirrup pots and variants, three have two spouts and three are trifids

(Greengo and Meighan 1976: Figs. 1, 2). In all six, the aperture is in the form of a small pot and the connecting tubes show the characteristic elbow. One specimen has a doughnut base similar to that of Figure 24 *c* herein; the diameter of the upper body is larger than that of the usual miniature in both these vessels.

Some details in decoration and shape differ from the specimens I know. The pot with the doughnut base evidently has small strips of clay applied vertically to the upper body. Another vessel is decorated with diagonal gashes on the upper body. Still another, whose lower body is cinctured, has similar gashes above and below the waist, as well as on the upper body. Punctation and/or gashes adorn the ridges of some of the miniature pots. Finally, one strange vessel is compound both horizontally and vertically. The two lower bodies are joined near the base and are further connected by a stirrup tube, atop which sits a small pot with the aperture. Decoration is modeled and asymmetrical. One body and the tube on the opposite side have raised ridges; the other lower body and its opposing tube, hobnail ornament. The typical sunburst motif is scarcely represented, although one vessel (Greengo and Meighan 1976: Fig. 3*b*) has a simplified approximation; near absence may reflect paucity of bules in the Apulco collection.

As part of the latter, Greengo and Meighan (1976: Fig. 4 *a, b*) illustrate two small effigy vessels which I think may not be Capacha. The faces are reminiscent of certain figurines from east Colima, known locally as *camotas.* The term is derived from *camote* (sweet potato) because of the form of the legs; although *camote* is a masculine noun, the moneros nonchalantly have made the figurine designation feminine, because representation is of females.

21. These do not look familiar. Mrs. Marguerite Ekholm suggests resemblance to specimens illustrated by Reichel-Dolmatoff (1957: Fig. 2, nos. 11, 15) for Momil.

22. There is suggestion of the neck-base channel in the ill-defined Chumbícuaro phase of Apatzingán (Michoacán). Undated, but relatively "early" in the local series, the phase is so scantily represented that vessel form is almost unknown, but decoration includes incision, grooving, punching, and rasping (Kelly 1947: 74, Pl. 12 *c–h*, Fig. 33; the profile of Fig. 33 *c* has a faint suggestion of the characteristic channel).

23. From survey and testing in the Huitzuco-Tepecoacuilco area of north Guerrero, Greengo (1967) reports "Preclassic" material but is not more specific.

24. Mr. Ric Reynolds reports having seen several Olmec-like figures in private collections in Colima and sent me the photograph of a seated stone specimen (personal communication, 6 August 1971). He is not certain of its authenticity; I am dubious but have not seen the figure.

25. A third human-effigy vessel attributed to Nayarit is of special interest, for its two tubes have the typical Capacha elbow or joint. For a photograph of this specimen, said to be in the Los Angeles County Museum of Natural History, I am indebted to Mr. Ralph Marshall (personal communication, 18 March 1978).

26. Miniature effigy bowls are attributed to Capacha (Fig. 14 *a–c*), and at least one tiny bowl, apparently not an effigy, comes from one of the tombs at El Opeño (Noguera [1942]: 584, Fig. 12 *b*). Such vessels are not cited among the ceramic resemblances because the phase of the specimen from El Opeño is unknown.

Several other traits are shared by Capacha and Opeño but are widespread in Mesoamerica in Preclassic times. One is the pinched-rim bowl (sometimes called bilobed or kidney shaped), which is reported also for Tlatilco and Chupícuaro, and even for Machalilla in Ecuador. Still another element is a series of small rim lobes or protuberances, with ornamental gashes—a trait shared by Tlatilco, although apparently not illustrated for Chupícuaro. Conceivably, these lobes may relate to the "flange-rim tradition" which Meggers and Evans (1969: Fig. 7) indicate was widespread in Mesoamerica, South America, and the Caribbean shortly before 1000 B.C.

27. Exceptions in the Colima area would be polished black wares, which are common in the eastern part of the state, as well as seals and stamps. Of these latter, the few I know from Colima are mostly without association. Field (1967: 38) reports 218 seals from Colima, of which 121 are attributed to El Chanal. Although this suggests a comparatively late appearance, circular stamps with a spiral motif, purchased in Coahuayana, allegedly were found with pottery which presumably is Preclassic.

28. More recently, the Tlatilco culture has been placed as "late Early Formative," dating from "about 1150 to 1000 B.C." (Grove and others 1976: 1205). Grennes-Ravitz and Coleman (1976: 198) state flatly that the Río Cuautla ceramics illustrated by Grove (1970: Figs. 5, 6) are not "post-Olmec," in which view Grove now concurs (see text above).

29. Ojochi is said to have fluting, gadrooning, and zoned red on brown, strongly suggestive of Niederberger's Nevada complex. Coe (1970: 22, 25) notes the presence of the bottle at the Tlatilco site and in the Ojochi and Bajío phases but discounts its importance because of apparent temporal discrepancy.

30. Nevertheless, one stirrup vessel attributed to Tlatilco is, in some respects, disconcertingly reminiscent of Capacha. It has been called to my attention through a thesis written by Ralph P. Marshall (*A distributional analysis of the stirrup-spout vessel and its possible relationship to the turquoise trade,* California State University, Los Angeles, 1978). Because of the recent date, the study is not listed in the bibliography. Reference is to a vessel figured some years ago in *Flor y canto* (1964: no. 150). The opening is through a small jar with raised swirls, a kind of decoration reminiscent of an unusual vessel, which I think may be Capacha (Eisleb 1971: no. 192; see also Greengo and Meighan 1976: Fig. 2 *c* for raised swirls). The two connecting tubes are elbowed. The lower body is not typical of either Tlatilco or Capacha; flattened globular, it lacks the characteristic Tlatilco carination and, moreover, has a human face in low relief on the side that is visible in the photograph. Also uncharacteristic, to judge from the color plate, are the purplish red wash and the apparently high polish, although these two traits are not unknown in Capacha association (Fig. 30 *b, bb*).

On the whole, Capacha resemblances outnumber the Tlatilco ones. The specimen is said to belong to the Museo Nacional in Mexico City; there I tried to locate the vessel but, at present, it is not on display or in storage. Conceivably it could form part of one of the loan collections sent out to affiliated local museums, but without the catalog number one cannot be sure. Similarly, without the identifying number, it is impossible to determine the recent history of the piece. A considerable part of the Tlatilco material in the Museo was acquired from private collections, assembled by the owners through purchase. Stirrup vessels are so rare in Mexico—and the one in question so obviously not Postclassic in style—that a dealer might well have guessed the specimen to be of Tlatilco manufacture and sold it as such. At present, we cannot be sure, but there is a good chance that the *Flor y canto* illustration may depict an unusual Capacha product.

31. Grove's most recent comment (1977: 635) on the archaeological situation in Morelos has his San Pablo B subphase include "Olmec stylistic motifs" in quantity, as well as "stirrup spout bottles, composite bottles, and belted bottles, often with resist decoration." These last-mentioned traits are, of course, characteristic elements of the Tlatilco style. Despite the time discrepancy (see note 28), Grove suggests that these several traits "seem to reflect interaction at this time with West Mexico."

32. The recent thesis of Ralph P. Marshall (see note 30) indicates that several trifids have been found in the Mississippi area. One specimen illustrated is attributed to the University of Arkansas collections. Identified as Hodges engraved, it is a fine two-storied vessel. The upper body is comparatively large, with a tallish spout; the lower one appears nearly globular; both have engraved curvilinear designs. The whole matter of Mesoamerican-Southeast relationships remains cloudy, and this is one of many puzzling specific resemblances—with occurrences widely separated spatially and temporally.

33. At present, close temporal comparisons are difficult. Chronological confusion is compounded by the several "corrections" and "adjustments" applied to radiocarbon dates. Even if the latter are plentiful, they are not comparable unless one knows precisely to what adjustments they have been subjected.

34. Thirty years ago, on ethnographic evidence culled from documentary sources and effigy representations, Kirchhoff (1948) called attention to the non-Mesoamerican flavor of west Mexico and pointed out resemblances to the northern Gulf coast, to some "regions of Peru and Ecuador," and particularly to the "Chimú culture."

It is significant that Weaver (1972: Maps 2–4), in reviewing the overall archaeological situation, does not place west Mexico within Mesoamerica until Postclassic times.

35. Bules are also of special interest because of possible function as a still. This novel interpretation was suggested tentatively by Dr. Joseph Needham and Dr. Lu Gwei-Djen, who saw some of the Capacha material in the Museo Nacional in the summer of 1977. They felt that the stirrup pots and the trifids with the aperture in the form of a vessel, as well as the bule described above as having upper and lower bodies separated by a perforated divider, could be used for distillation provided a cover were placed over the vessel mouth, with a small receptacle beneath to catch the drip. They suggested that such an arrangement—in principle, although not in specifics—would parallel Chinese equipment, roughly contemporaneous with Capacha, as well as the modern Huichol still.

36. Spence and Weigand (1968) mention obsidian workshops at Carmelita, in the Atoyac area (Jalisco), not far from Colima. These appear to be "largely in the Classic period, and may even extend back into the late Preclassic." As yet, the source of the obsidian has not been determined (personal communication from Dr. Michael Spence, 11 September 1974).

37. Obviously, this comment is not intended to suggest cacao cultivation in Colima during the second millennium before our era. I have no idea of the date it appeared in west Mexico, but indubitably it was pre-Conquest. It is strange, however, that the one early report for the stretch from Autlán (Jalisco) to southern Nayarit scarcely mentions cacao except for two pueblos not far from Tepic ([Cortés] 1937: 564). This seems particularly odd because the document contains a wealth of data concerning agriculture, irrigation, and local trade.

In the mid-sixteen century, cacao was of enormous economic importance in Colima (Lebrón 1951: 114–15). Lebrón (1951: 12) states flatly that the large-scale exploitation there was a matter of Spanish enterprise and was responsible, to a considerable extent, for the early disappearance of the native population, because the best cultivable and irrigable lands were devoted to the *heredades,* as cacao plantings were called.

Offhand, I do not recall representations of cacao fruit on any Comala-phase vessels, but years ago, Sra. María Ahumada de Gómez showed me a few charred "beans" she had found archaeologically and which she believed were cacao. Unfortunately, they have not been examined by a specialist, and their ceramic association is unknown.

REFERENCES

The present study was completed late in February of 1975. Owing to delay in publication, a number of papers then cited in manuscript now have been published. As of March, 1978, bibliographic entries for them have been altered and page references inserted in the text (for Grennes-Ravitz and Coleman 1976; Harbottle 1975; Kelly 1978; Mountjoy and Weigand 1975; Niederberger 1976; and von Winning 1976). Moreover, several new works have appeared (Greengo and Meighan 1976; Grove 1977; Grove and others 1976; and Meighan 1976). The text proper has not been rewritten to include the new data but the titles have been added to the bibliography and pertinent content reported in the End notes.

Acosta, Jorge R.
1956– Interpretación de Algunos de los Datos Obtenidos en
1957 Tula Relativos a la Epoca Tolteca. VI Mesa Redonda de la Sociedad Mexicana de Antropología. *Revista Mexicana de Estudios Antropológicos* 14(2): 75–110. México.

Alsberg, John L., and Rodolfo Petschek
1968 *The Evolution of Artistic Form: Ancient Sculpture from Western Mexico.* Berkeley: Nicole Gallery.

Anónimo
1968 Noticias de los Museos. *INAH, Boletín* 31: 51–57. México.

Aveleyra Arroyo de Anda, Luis
1964 *Obras Selectas del Arte Prehispánico: Adquisiciones Recientes.* Consejo para la Planeación e Instalación del Museo Nacional de Antropología. México: [Secretaría de Educación Pública].

Bell, Betty
1971 Archaelogy of Nayarit, Jalisco, and Colima. Archaeology of Northern Mesoamerica, Part 2, edited by Gordon F. Ekholm and Ignacio Bernal, pp. 694–753. *Handbook of Middle American Indians,* Vol. 11, general editor Robert Wauchope. Austin: University of Texas Press.

Bell, Betty (editor)
1974 *The Archaeology of West Mexico.* Ajijic, Jalisco: West Mexican Society for Advanced Study.

Bennett, Wendell C.
1944 The North Highlands of Peru: Excavations in the Callejón de Huaylas and at Chavín de Huántar. *American Museum of Natural History, Anthropological Papers* 39(1): 1–114.
1946 The Archeology of the Central Andes. Handbook of South American Indians 2:61–147. *Bureau of American Ethnology Bulletin* 143. Washington.

Benson, Elizabeth P. (editor)
1971 Dumbarton Oaks Conference on Chavín, October, 1968. Washington: Dumbarton Oaks Research Library and Collection.

Berger, R., and W. F. Libby
1967 UCLA Radiocarbon Dates VI. *Radiocarbon* 9: 477–504.
1970 UCLA Radiocarbon Dates X. Cited by Taylor 1970: 168 as in press in *Radiocarbon* 12(1). I have been unable to locate the article in question.

Brainerd, George W.
1949 A Stirrup Pot from Lower California. *Masterkey* 23: 5–8. Los Angeles: Southwest Museum.

Brand, Donald D., and others
1960 *Coalcomán and Motines del Oro: an Ex-Distrito of Michoacán, Mexico.* Published for the Institute of Latin American Studies, University of Texas. The Hague: Martinus Nijhoff.

Braniff, Beatriz
1974 Oscilación de la Frontera Septentrional Mesoamericana. In *The Archaeology of West Mexico,* edited by Betty Bell, pp. 40–50. Ajijic, Jalisco: West Mexican Society for Advanced Study.

Bray, Warwick
1970 Ancient Mesoamericana: Precolumbian Mexican and Maya Art. An Exhibition of Material from Private Collections in Great Britain. [Birmingham]: Birmingham Museum and Art Gallery.

Brush, Charles F.
1965 Pox Pottery: Earliest Identified Mexican Ceramic. *Science* 149(3680): 194–95.
1969 A Contribution to the Archeology of Coastal Guerrero, Mexico. Ph.D. Dissertation, Faculty of Political Science, Columbia University. Ann Arbor: University Microfilms.

Carrasco, Pedro
1971 The Peoples of Central Mexico and Their Historical Traditions. Archaeology of Northern Mesoamerica, Part 2, edited by Gordon F. Ekholm and Ignacio Bernal, pp. 459–73. *Handbook of Middle American Indians,* Vol. 11, general editor Robert Wauchope. Austin: University of Texas Press.

Castellanos, Aniceto
1952 Riqueza y Primor de la Arqueología Colimense. In Moreno 1952: pp. 31–47. Mexico.

Cerámica Jalisciense
1964 Joyas del Museo de Arqueología del Occidente de México. Guadalajara, Jalisco: Instituto Nacional de Antropología e Historia, Instituto Jalisciense de Antropología e Historia.

Cervantes de Salazar, Francisco
1914 Crónica de Nueva España. Vol. I. *Papeles de Nueva España,* compilados y publicados por Francisco del Paso y Trancoso. Tercera serie. Madrid.

Chadwick, Robert
1971a Postclassic Pottery of the Central Valleys. Archaeology of Northern Mesoamerica, Part 1, edited by Gordon F. Ekholm and Ignacio Bernal, pp. 228–57. *Handbook of Middle American Indians,* Vol. 10, general editor Robert Wauchope. Austin: University of Texas Press.
1971b Archaeological Synthesis of Michoacán and Adjacent Regions. Archaeology of Northern Mesoamerica, Part 2, edited by Gordon F. Ekholm and Ignacio Bernal, pp. 657–93. *Handbook of Middle American Indians,* Vol. 11, general editor Robert Wauchope. Austin: University of Texas Press.

Cobean, Robert
1974 The Ceramics of Tula. In: Diehl 1974: 32–41.

Coe, Michael D.
1961 La Victoria. *Papers of the Peabody Museum of American Archaeology and Ethnology, Harvard University,* 53.
1963 Olmec and Chavín: Rejoinder to Lanning. *American Antiquity* 29(1): 101–104.
1970 The Archaeological Sequence at San Lorenzo Tenochtitlan, Veracruz, Mexico. *Contributions of the University of California Archaeological Research Facility* 8: 21–34. Berkeley: University of California Press.

Coe, Michael D., and Kent V. Flannery
1967 Early Cultures and Human Ecology in South Coastal Guatemala. *Smithsonian Contributions to Anthropology 3.* Washington: Smithsonian Institution Press.

Corona Núñez, José
1960a Arqueología: Occidente de México. Jalisco en el Arte. Guadalajara, Jalisco: Planeación y Promoción, S. A.
1960b Danza de Xipe-Totec. *Eco* 1, Enero de 1960. Guadalajara, Jalisco.
1960c Investigación Arqueológica Superficial Hecha en el Sur de Michoacán. In: Brand and others 1960: 366–403 (Appendix).

[Cortés, Francisco]
1937 Nuño de Guzmán contra Hernán Cortés sobre los Descubrimientos y Conquistas en Jalisco y Tepic, 1531. *Boletín del Archivo General de la Nación* 8(4): 541–76.

Crane, H. R., and James B. Griffin
1972 University of Michigan Radiocarbon Dates XIV. *Radiocarbon* 14(1): 155–94.

Davies, Claude Nigel
1973 Los Mexicas: Primeros Pasos hacia el Imperio. Instituto de Investigaciones Históricas. *Serie de Cultura Náhuatl, Monografías* 14. Mexico: Universidad Nacional Autónoma de México.

Dibble, Charles E., and Arthur J. O. Anderson
1963 Florentine Codex. General History of the Things of New Spain, by Fray Bernardino de Sahagún. Book 11, Earthly Things. *Monographs of The School of American Research and The Museum of New Mexico* 14, Part XII. Santa Fe.

Diehl, Richard A. (editor)
1974 Studies of Ancient Tollan: A Report of the University of Missouri Tula Archaeological Project. *University of Missouri Monographs in Anthropology* 1. Columbia.

Dixon, Keith A.
1959 Ceramics from Two Preclassic Periods at Chiapa de Corzo, Chiapas, Mexico. *Papers of the New World Archaeological Foundation* 5. Orinda, California.

1964 The Acceptance and Persistence of Ring Vessels and Stirrup Spout-handles in the Southwest. *American Antiquity* 29(4): 455–60.

Dumond, D. E., and Florencia Müller
1972 Classic to Postclassic in Highland Central Mexico. *Science* 175(4027): 1208–15.

Easby, Elizabeth Kennedy, and John F. Scott
1970 *Before Cortés: Sculpture of Middle America.* A Centennial Exhibition at the Metropolitan Museum of Art. Distributed by New York Graphic Society.

Eisleb, Dieter
1971 Westmexikanische Keramik. Neue Folge 24, Abteilung Amerikanische Archäologie II. Veröffentlichungen, Museums für Völkerkunde Berlin. Berlin: Staatliche Museen Preussischer Kulturbesitz.

Estrada, Emilio
1957 Prehistoria de Manabí. *Publicatión del Museo Victor Emilio Estrada* 4. Guayaquil. [Reference, courtesy of Dr. Allison Paulsen.]

Evans, Clifford, and Betty J. Meggers
1960 Archeological Investigations in British Guiana. Smithsonian Institution, *Bureau of American Ethnology Bulletin* 177. Washington.
1966 Mesoamerica and Ecuador. Archaeological Frontiers and External Connections, edited by Gordon F. Ekholm and Gordon R. Willey, pp. 243–64. *Handbook of Middle American Indians,* Vol. 4, general editor Robert Wauchope. Austin: University of Texas Press.

Field, Frederick V.
1967 Thoughts on the Meaning and Use of Pre-Hispanic Mexican Sellos. *Studies in Pre-Columbian Art and Archaeology* 3. Washington: Dumbarton Oaks.

Fish, Louise
1974 Figurines with Up-tilted Noses from Colima, Mexico. In: Bell 1974: 212–14.

Flor y canto del arte prehispánicó de México
1964 Fondo Editorial de la Plástica Mexicana. México: Banco de Comercio Exterior, S.A.

Ford, James A.
1969 A Comparison of Formative Cultures in the Americas. *Smithsonian Contributions to Anthropology* 11. Washington: Smithsonian Institution Press.

Furst, Peter T.
1965a West Mexican Tomb Sculpture as Evidence for Shamanism in Prehispanic Mesoamerica. Separata de *Antropologica* 15: 29–60. Instituto Caribe de Antropología y Sociología, Caracas. Latin American Center, Reprint 5. Los Angeles: University of California Press.
1965b West Mexico, the Caribbean, and Northern South America: Some Problems in New World Interrelationships. *Anthropológica* 14: 1–37. Caracas.

Gifford, Edward W.
1950 Surface Archaeology of Ixtlán del Río, Nayarit. *University of California Publications in American Archaeology and Ethonology* 43(2): i–viii, 183–302. Berkeley and Los Angeles: University of California Press.

Glassow, Michael A.
1967 The Ceramics of Huistla, a West Mexican Site in the Municipality of Etzatlán, Jalisco. *American Antiquity* 32(1): 64–83.

Green, Dee F., and Gareth W. Lowe
1967 Altamira and Padre Piedra, Early Preclassic Sites in Chiapas, Mexico. *Papers of the New World Archaeological Foundation* 20, Publication 15. Provo: Brigham Young University Press.

Greengo, Robert E.
1967 Reconocimiento Arqueológico en el Noroeste [sic: Nordeste?] de Guerrero. *INAH, Boletín* 29: 6–10. México.

Greengo, Robert E., and Clement W. Meighan
1976 Additional Perspective on the Capacha Complex of Western Mexico. Institute of Archaeology I(5): 15–23. Los Angeles: University of California.

Grennes-Ravitz, Ronald A.
1974 The Olmec Presence at Iglesia Vieja, Morelos. In: Hammond 1974: 99–108.

Grennes-Ravitz, Ronald A., and G. H. Coleman
1976 The Quintessential Role of Olmec in the Central Highlands of Mexico: a Refutation. *American Antiquity* 41(2): 196–206.

Gross, Daniel R. (editor)
1973 *Peoples and Cultures of Native South America: an Anthropological Reader*. Published for the American Museum of Natural History. Garden City, New York: Doubleday/The Natural History Press.

Grosscup, Gordon L.
1963 A Sequence of Figurines from West Mexico. *American Antiquity* 26(3): 390–406.

Grove, David C.
1970 The San Pablo Pantheon Mound: a Middle Preclassic Site in Morelos, Mexico. *American Antiquity* 35(1): 62–73.
1971 The Mesoamerican Formative and South American Influences. Primer Simposio de Correlaciones Antropológicas Andino-Mesoamericano. Salinas, Ecuador. Mimeographed.
1974a The Highland Olmec Manifestation: a Consideration of What It Is and Isn't. In: Hammond 1974: 109–28.
1974b San Pablo, Nexpa, and the Early Formative Archaeology of Morelos, Mexico. *Vanderbilt University Publications in Anthropology* 12. Nashville.
1977 The Central Mexican Preclassic: is there really disagreement? *American Antiquity* 42(4): 634–37.

Grove, David C., Kenneth G. Hirth, David E. Bugé, Ann M. Cyphers
1976 Settlement and Cultural Development at Chalcatzingo. *Science* 192: 1203–10.

Hammond, Norman (editor)
1974 *Mesoamerican Archaeology: New Approaches*. Austin: University of Texas Press.

Harbottle, Garman
1975 Activation Analysis Study of Ceramics from the Capacha (Colima) and Opeño (Michoacán) Phases of West Mexico. *American Antiquity* 40(4): 453–58.

Healy, Paul F.
1974 The Cuyamel Caves: Preclassic Sites in Northeast Honduras. *American Antiquity* 39(3): 435–47.

Heizer, Robert F., and John A. Graham (editors)
1971 Observations on the Emergence of Civilization in Mesoamerica. *Contributions of the University of California Archaeological Research Facility* 11. Berkeley: University of California Press.

Heyden, Doris
1970 Nueva Interpretación de las Figuras Sonrientes, Señalada por las Fuentes Históricas. *Tlalocan* 6(2): 159–62. México.

Kan, Michael, Clement Meighan, and H. B. Nicholson
1970 *Sculpture of Ancient West Mexico: Nayarit, Jalisco, Colima: the Proctor Stafford Collection*. Los Angeles: Los Angeles County Museum of Art.

Kelley, J. Charles
1974 Speculations on the Culture History of Northwestern Mesoamerica. In: Bell 1974: 19–39.

Kelley, J. Charles, and Ellen Abbott
1966 The Cultural Sequence on the North Central Frontier of Mesoamerica. *Actas y Memorias del XXXVI Congreso Internacional de Americanistas*, 1: 325–44. Seville.

Kelley, J. Charles, and Ellen Abbott Kelley
1971 An Introduction to the Ceramics of the Chalchihuites Culture of Zacatecas and Durango, Mexico. I, The Decorated Wares. University Museum, *Mesoamerican Studies* 5. Carbondale: Southern Illinois University.

Kelly, Isabel
[1944] West Mexico and the Hohokam. In: El Norte de México y el Sur de Estados Unidos. *Tercera Reunión de Mesa Redonda sobre Problemas Antropológicos de México y Centro America*, pp. 206–22. Mexico: Sociedad Mexicana de Antropología. 1943.
1945a Excavations at Culiacán, Sinaloa. *Ibero-Americana* 25. Berkeley and Los Angeles: University of California Press.
1945b The Archaeology of the Autlán-Tuxcacuesco Area of Jalisco. I: The Autlán Zone. *Ibero-American* 26. Berkeley and Los Angeles: University of California Press.
1947 Excavations at Apatzingán, Michoacán. *Viking Fund Publications in Anthropology* 7. New York.
1948 Ceramic Provinces of Northwest Mexico. In: El Occidente de México. *Cuarta Reunión de Mesa Redonda sobre Problemas Antropológicos de México y Centro América*, pp. 55–71, Fig. IX. Mexico: Sociedad Mexicana de Antropología. 1946 (1947).
1949 The Archaeology of the Autlán-Tuxcacuesco Area of Jalisco. II: The Tuxcacuesco-Zapotitlán Zone. *Ibero-American* 27. Berkeley and Los Angeles: University of California Press.
1972 Vasijas de Colima con Boca de Estribo. *INAH, Boletín* 42: 26–31. Diciembre 1970. México.
1974 Stirrup Pots from Colima: Some Implications. In: Bell 1974: 206–11.
1978 Seven Colima Tombs: An interpretation of ceramic content. Studies in Mesoamerica III. *Contributions of the University of California Archaeological Research Facility* 36: 1–26. Berkeley: University of California Press.
n.d. A Surface Survey of the Sayula-Zacoalco Basins of Jalisco. Ms.

Kelly, Isabel, and Angel Palerm
1952 The Tajín Totonac. Part 1. History, Subsistence, Shelter, and Technology. Smithsonian Institution, *Institute of Social Anthropology* 13. Washington.

Kelly, Isabel, and Beatriz B. de Torres
1966 Una Relación Cerámica entre Occidente y la Mesa Central. *INAH, Boletín* 23: 26–27. México.

Kidder, Alfred V., Jesse D. Jennings, and Edwin M. Shook
1946 Excavations at Kaminaljuyu, Guatemala. *Carnegie Institution of Washington Publication* 561. Washington.

Kirchhoff, Paul
1948 Etnografía Antigua. In: El Occidente de México, pp. 134–36.
1961 Das Toltekenreich und sein Untergang. *Saeculum* 12(3): 248–65. Munich.
n.d. El Imperio Tolteca y su Ocaso. [Spanish translation of the preceding]. Typescript.

Lanning, Edward P.
  1967 *Peru before the Incas.* Englewood Cliffs: Prentice-Hall.
  1968 Investigaciones Arqueológicas en la Península de Santa Elena, Ecuador. Informe entregado a la Casa de la Cultura Ecuatoriana. Mimeographed.

Lebrón de Quiñones, Lorenzo
  1951 *Relación Breve y Sumaria de la Visita Hecha por el Lic. Lorenzo Lebrón de Quiñones, Oidor del Nuevo Reino de Galicia, por Mandado de su Alteza.* Nota Introductiva del Dr. Rubén Villaseñor Bordes. Ediciones de la Junta Auxiliar Jalisciense de la Sociedad Mexicana de Geografía y Estadística 9: 1–123. Guadalajara, Jalisco.

Linné, S.
  1934 Archaeological Researches at Teotihuacán, Mexico. *The Ethnographical Museum of Sweden,* n.s. 1. Stockholm.

Lister, Robert H.
  1949 Excavations at Cojumatlán, Michoacán, México. *University of New Mexico Publications in Anthropology* 5. Albuquerque: University of New Mexico Press.
  1971 Archaeological Synthesis of Guerrero. Archaeology of Northern Mesoamerica, Part 2, edited by Gordon F. Ekholm and Ignacio Bernal, pp. 619–31. *Handbook of Middle American Indians,* Vol. 11, general editor Robert Wauchope. Austin: University of Texas Press.

Litvak King, Jaime
  1968 Excavaciones de Rescate en la Presa de la Villita. *INAH, Boletín* 31: 28–30. México.

Long, Stanley Vernon
  1966 Archaeology of the Municipio of Etzatlán, Jalisco. Ph.D. dissertation, University of California, Los Angeles. Ann Arbor: University Microfilms.
  1967 Formas y Distribución de Tumbas de Pozo con Cámara Lateral. Separata de *Razón y Fábula,* Revista de la Universidad de los Andes.

Long, Stanley Vernon, and R. E. Taylor
  1966 Suggested Revision for West Mexican Archeological Sequences. *Science* 154(3755): 1456–59.

Lorenzo, José Luis
  1965 Tlatilco: los Artefactos. III. *Investigaciones* 7. Mexico: Instituto Nacional de Antropología e Historia.

Lumbreras, Luis Guillermo
  1971 Towards a Re-evaluation of Chavín. In: Benson 1971: 1–28.

McBride, Harold W., and Diego W. Delgado
  n.d. Cerámica de Estilo Teotihuacano en Colima. Mimeographed.

MacNeish, Richard S.
  1954 An Early Archaeological Site near Pánuco, Vera Cruz. *Transactions of the American Philosophical Society* n.s. 44(5): 357–641. Philadelphia.

Matos Moctezuma, Eduardo, and Isabel Kelly
  1974 Una Vasija que Sugiere Relaciones entre Teotihuacán y Colima. In: Bell 1974: 202–205.

Médioni, Gilbert, and Marie-Thérèse Pinto
  1941 *Art in Ancient Mexico.* New York: Oxford University Press.

Meggers, Betty J., and Clifford Evans
  1969 Speculations of Early Pottery Diffusion Routes between South and Middle America. *Biotropica* 1(1): 20–27.

Meggers, Betty J., Clifford Evans, and Emilio Estrada
  1965 Early Formative Period of Coastal Ecuador: the Valdivia and Machalilla Phases. *Smithsonian Contributions to Anthropology* 1. Washington: Smithsonian Institution Press.

Meighan, Clement W.
  1972 Archaeology of the Morett Site, Colima. *University of California Publications in Anthropology* 7. Berkeley and Los Angeles: University of California Press.
  1974 Prehistory of West Mexico. *Science* 184(4143): 1254–61.

Meighan, Clement W. (editor)
  1976 The Archaeology of Amapa, Nayarit. Institute of Archaeology, *Monumenta Arqueologica,* Vol. 2. Los Angeles: University of California Press.

Meighan, Clement W., Frank J. Findlow, and Suzanne P. De Atley (editors)
  1974 A Compendium of the Obsidian Determinations Made at the UCLA Obsidian Hydration Laboratory. *Archaeological Survey Monograph* 3. Institute of Archaeology, Archaeological Survey. Los Angeles: University of California Press. Photo offset.

Meighan, Clement W., and Leonard J. Foote
  1968 Excavations at Tizapan El Alto, Jalisco. *Latin American Studies* 11. Latin American Center. Los Angeles: University of California Press.

Meighan, Clement W., and Robert E. Greengo
  1974 Additional Perspective on the "Capacha" complex in West Mexico. Abstract of paper presented at the 39th annual meeting of the Society for American Archaeology. Washington.

Menzel, Dorothy
  1973 Archaism and Revival on the South Coast of Peru. In: Gross 1973: 19–25.

Messmacher, Miguel
  1966 Colima. Instituto Nacional de Antropología e Historia, *Colección de Libros de Arte* 1. México: Secretaría de Educación Pública.

Millon, René
  1973 The Teotihuacán Map. Part 1: text. In: Millon (editor) 1973: i–xvi, 1–64.

Millon, René (editor)
  1973 Urbanization at Teotihuacán, Mexico, Vol. 1. The Dan Danciger Publication Series. Austin and London: University of Texas Press.

Millon, René, Bruce Drewitt, and James A. Bennyhoff
  1965 The Pyramid of the Sun at Teotihuacán: 1959 Investigations. *Transactions of the American Philosophical Society* n.s. 55(6): 1–93. Philadelphia.

Moreno, Daniel
  1952 *Colliman* [1]. *Ensayo Enciclopédico.* México.

Mountjoy, Joseph
  1970 La Sucesión Cultural en San Blas. *INAH, Boletín* 39: 41–49. México.

Mountjoy, Joseph B., and Phil C. Weigand
  1975 The Prehispanic Settlement Zone at Teuchitlán, Jalisco. *Actas del XLI Congreso Internacional de Americanistas* I: 353–63. Mexico.

Munizaga, Juan R.
  1965 Skeletal Remains from Sites of Valdivia and Machalilla Phases. In: Meggers, Evans, and Estrada 1965: 219–34 (Appendix 2).

Museo Diego Rivera-Anahuacalli
  1968 México, Comité Organizador de los Juegos de la XIX Olimpiada.

Natalie Wood Collection of Pre-Columbian Ceramics at UCLA
  1969 *Occasional Papers of the Museum and Laboratories of*

Ethnic Arts and Technology 1. Los Angeles: University of California Press.

Niederberger, Christine
1976 Zohapilco: Cinco Milenios de Ocupación Humana en un Sitio Lacustre de la Cuenca de México. Instituto Nacional de Antropología e Historia, Departmento de Prehistoria, *Colección Científica* 30. México.

Noguera, Eduardo
[1942] Exploraciones en "El Opeño", Michoacán. *Actas de la Primera Sesión del Vigesimoséptimo Congreso Internacional de Americanistas* 1: 574–86. 1939. México: Instituto Nacional de Anthropología e Historia, Secretaría de Educación Pública.

Norte de México y el Sur de Estados Unidos, El
[1944] *Tercera Reunión de Mesa Redonda sobre Problemas Antropológicos de México y Centro América.* México: Sociedad Mexicana de Antropología. 1943.

Occidente de México, El
1948 *Cuarta Reunión de Mesa Redonda sobre Problemas Antropológicos de México y Centro América.* México: Sociedad Mexicana de Antropología. 1946 (1947).

Oliveros, José Arturo
1970 Excavación de dos Tumbas en El Opeño, Michoacán. Tesis profesional, Escuela Nacional de Antropología e Historia, y Universidad Nacional Autónoma de México. [México.] Mimeographed.
1974 Nuevas Exploraciones en El Opeño, Michoacán. In: Bell 1974: 182–201.

Parsons, Lee A.
1963 A Doughnut-shaped Vessel from Kaminaljuyú, with a Distributional Analysis of this Unusual Form. *American Antiquity* 28(3): 386–92.

Paso y Troncoso, Francisco del (editor)
1905 *Papeles de Nueva España, Segunda Serie,* Geografía y Estadística 1, Suma de Visitas de Pueblos por Orden Alfabético. Madrid.

Paulsen, Allison C., and Eugene J. McDougle
1974 The Machalilla and Engoroy Occupations of the Santa Elena Peninsula in South Coastal Ecuador. Paper presented at the 39th annual meeting of the Society for American Archaeology, Washington, D.C. Mimeographed.

Pendergast, David M.
1962a Metal Artifacts from Amapa, Nayarit, Mexico. *American Antiquity* 27(3): 370–79.
1962b Metal Artifacts in Prehispanic Mesoamerica. *American Antiquity* 27(4): 520–45.

Phillips, Philip, James A. Ford, and James B. Griffin
1951 Archaeological Survey on the Lower Mississippi Alluvial Valley, 1940–1947. *Papers of the Peabody Museum of American Archaeology and Ethnology, Harvard University,* 25.

Piña Chan, Román
1958 Tlatilco I. *Investigaciones* 1. México: Instituto Nacional de Antropología e Historia.

Ponce, Alonso
1873 *Relación Breve y Verdadera de Algunas Cosas de las Muchas que Sucedieron al Padre Fray Alonso Ponce en las Provincias de la Nueva España. Escrita por Dos Religiosos, sus Compañeros. Vol. II. Madrid.*

Porter, Muriel Noé
1953 Tlatilco and the Pre-Classic Cultures of the New World. *Viking Fund Publications in Anthropology* 19. New York.

1956 Excavations at Chupícuaro, Guanajuato, Mexico. *Transactions of the American Philosophical Society* n.s. 46(5): 513–637. Philadelphia.

Ralph, E. K., H. N. Michael, and M. C. Han
1973 Radiocarbon Dates and Reality. *MASCA Newsletter* 9(1): 1–18. Applied Science Center for Archaeology, The University Museum. Philadelphia: University of Pennsylvania.

Rattray, Evelyn Childs
1966 An Archeological and Stylistic Study of Coyotlatelco Pottery. Teotihuacán and After. *Mesoamerican Notes* 7–8: 87–193. México: University of the Americas.

Reichel-Dolmatoff, Gerardo
1957 Momil: a Formative Sequence from the Sinú Valley, Colombia. *American Antiquity* 22(3): 226–34.

Reichel-Dolmatoff, Gerardo, and Alicia Reichel-Dolmatoff
1956 Momil: Excavaciones en El Sinú. *Revista Colombiana de Antropología* V(4): 109–333.

Relación de Ameca
1878 Discripcion Hecha por el Ilustre Señor Antonio de Leyva, Alcalde Mayor por S. M. del Pueblo de Ameca, Año de 1579. *Noticias Varias de Nueva Galicia, Intendencia de Guadalajara,* pp. 233–82. Guadalajara, [Jalisco].

Reyes Mazzoni, Roberto, and Vito Véliz
1974 La Cerámica de Cuyamel. *Revista de la Universidad* [Nacional Autónoma de Honduras]. Etapa V(8): 3–26. [Tegucigalpa, Honduras]: Editorial Universitaria.

Reyna, Rosa María
1971 Las Figurillas Preclásicas. Tesis profesional, Escuela Nacional de Antropología e Historia, y Universidad Nacional Autónoma de México. México. Mimeographed.

Romano, Arturo
1963 Exploraciones en Tlatilco. *INAH, Boletín* 14: 11–14. México.
1972 Craneo Precerámico con Deformación Intencional. *INAH, Boletín,* Epoca II, 1: 35–36. México.

Rosado Ojeda, Vladimiro
1948 Interpretación de la Grada Jeroglífica del Chanal, Colima. In: *El Occidente de México,* pp. 72–73, Fig. X.

Sauer, Carl
1934 The Distribution of Aboriginal Tribes and Languages in Northwestern Mexico. *Ibero-Americana* 5. Berkeley: University of California Press.
1948 Colima of New Spain in the Sixteenth Century. *Ibero-Americana* 29. Berkeley and Los Angeles: University of California Press.

Sauer, Carl, and Donald Brand
1932 Aztatlán. *Ibero-Americana* 1. Berkeley: University of California Press.

Schöndube Baumbach, Otto
1973– Tamazula-Tuxpan-Zapotlán: Pueblos de la Frontera
1974 Septentrional de la Antigua Colima I, II. Tesis, ENAH. [México.]
1974 Deidades Prehispánicas en el Area de Tamazula-Tuxpan-Zapotlán en el Estado de Jalisco. In: Bell 1974: 168–81.

Spence, Michael W., and Phil C. Weigand
1968 Some Patterns of Obsidian Exploitation and Trade in Northern Mesoamerica. Paper presented to the 33rd annual meeting of the Society for American Archeology, Santa Fe, 1968. Mimeographed.

Tamayo, Rufino
  1973   *Arte Prehispánico de México.* Colección Rufino Tamayo. México: Ediciones Galería de Arte Misrachi.
Taylor, R. E.
  1970   The Shaft Tombs of Western Mexico: Problems in the Interpretation of Religious Function in Nonhistoric Archaeological Context. *American Antiquity* 35(2): 160–69.
Taylor, R. E., Rainer Berger, C. W. Meighan, and H. B. Nicholson
  1969   West Mexican Radiocarbon Dates of Archaeologic Significance. In: Natalie Wood Collection 1969: 17–30.
Thompson, J. Eric S.
  1970   *Maya History and Religion.* Norman: University of Oklahoma Press.
Tolstoy, Paul
  1958   Surface Survey of the Northern Valley of Mexico: the Classic and Post-Classic Periods. *Transactions of the American Philosophical Society* n.s. 48(5): 1–101. Philadelphia.
  1971   Recent Research into the Early Preclassic of the Central Highlands. *Contributions of the University of California Archaeological Research Facility* 11: 25–28. Berkeley: University of California Press.
  1973   The Archaeological Chronology of Western Mesoamerica before 900 A.D. 1970; second revision, June, 1973. Mimeographed.
Tolstoy, Paul, and Louise I. Paradis
  1970   Early and Middle Preclassic Culture in the Basin of Mexico. *Science* 167(3917): 344–51.
  1971   Early and Middle Preclassic Culture in the Basin of Mexico. *Contributions of the University of California Archaeological Research Facility* 11: 7–25. Berkeley: University of California Press.

Vaillant, George C.
  1930   Excavations at Zacatenco. *American Museum of Natural History, Anthropological Papers* 32(I): 1–197.
  1931   Excavations at Ticomán. *American Museum of Natural History, Anthropological Papers* 32(II): 199–439.
  1935   Excavations at El Arbolillo. *American Museum of Natural History, Anthropological Papers* 35(II): 135–279.
Vaillant, Suzannah B., and George C. Vaillant
  1934   Excavations at Gualupita. *American Museum of Natural History, Anthropological Papers* 35(I): 1–135.
von Winning, Hasso
  1969   *Pre-Columbian Art of Mexico and Central America.* Selection of plates by Alfred Stendahl. London: Thames and Hudson.
  1976   Escenas Rituales en la Cerámica Policroma de Nayarit. *Actas del XLI Congreso Internacional de Americanistas* II: 387–400. México.
Weaver, Muriel Porter
  1967   Tlapacoya Pottery in the Museum Collections. *Indian Notes and Monographs. Miscellaneous Series* 56. New York: Museum of the American Indian, Heye Foundation.
  1972   *The Aztecs, Maya, and Their Predecessors.* New York and London: Seminar Press.
Weigand, Phil C.
  1974   The Ahualulco Site and the Shaft-tomb Complex of the Etzatlán Area. In: Bell 1974: 120–31.
Weitlaner, Robert J., and R. H. Barlow
  1944   Expeditions in Western Guerrero: the Weitlaner Party. Spring, 1944. *Tlalocan* 1(4): 364–75.
Willey, Gordon R.
  1971   *An Introduction to American Archaeology,* Vol. 2. Englewood Cliffs: Prentice-Hall.